Restoring farmers' seed systems in disaster situations

Seed and Plant Genetic Resources Services
Plant Production and Protection Division

Published by arrangement with the
Food and Agriculture Organization of
the United Nations
by
Daya Publishing House®
A Division of
Astral International Pvt. Ltd.
New Delhi - 110 002

"This book was originally published by the Food and Agriculture Organization of the United Nations (FAO) as FAO Plant Production and Protection Paper No. 150".

First Indian Edition, 2005
Reprinted, 2016
ISBN: 978-81-7035-395-9

Published by : **Daya Publishing House®**
A Division of
Astral International Pvt. Ltd.
– ISO 9001:2008 Certified Company –
4760-61/23, Ansari Road, Darya Ganj
New Delhi-110 002
Ph. 011-43549197, 23278134
E-mail: info@astralint.com
Website: www.astralint.com

Printed at : **Salasar Imaging System**

TABLE OF CONTENTS

PREFACE

Planned conservation and utilization of local crop genetic resources is of prime importance for food security, as it permits small-scale farmers of developing countries to have permanent access to seed and planting materials adapted to their region. Conservation of and access to these resources are most important in disaster-prone areas, as farmers in these regions often lose all of their seed stocks during disasters. Furthermore, disasters cause the loss of local varieties or genotyes, hence leading to the erosion of genetic diversity.

Following these calamities, seed of adapted varieties is the major input requested by the agricultural sector. Usually, it is difficult to provide local communities with seed and planting material of well adapted improved varieties or local genotypes, when this indigenous genetic variability is jeopardized or lost. Distribution of grain to be used for seed, instead of good quality seed, produces low yields and increases risk of introducing pests and diseases. Furthermore, the introduction of foreign varieties, which are not adapted to local conditions, leads to low yields and food quality which may not always suit local tastes and preferences. Additionally, the introduction of such foreign varieties induces uncontrolled crossings with local genotypes adapted to the region. This "genetic pollution" is responsible for the definitive loss of local genetic resources and, as a result, of the disruption of the sustainable agricultural system of the region concerned.

In order to better assist countries regularly affected by calamities, it is necessary to implement effective strategies and policies of seed security which are also aimed at protecting local crop genetic diversity of the region and subsequently to reproduce adapted genotypes to rehabilitate agricultural systems affected by disasters.

This workshop was organised under FAO's mandate to facilitate the implementation of the Global Plan of Action (GPA) for the conservation and sustainable utilisation of crop genetic resources for food and agriculture, adopted by the International Technical Conference on Plant Genetic Resources, held in Leipzig, Germany in June, 1996.

ACKNOWLEDGEMENTS

The Seed and Plant Genetic Resources Service of the Food and Agriculture Organization of the United Nations is greatly indebted to the Government of Norway for providing the funding to convene this International Workshop on developing institutional agreements and capacity to assist farmers in disaster situations to restore agricultural systems and seed security activities (Project GCP/INT/660/NOR) and to the many participants who made this Workshop a success.

Special thanks are also addressed to all those who assisted in the implementation of the Workshop by providing information and advice during the initial Planning Meeting convened in February 1998 which provided the background work to prepare for this Workshop.

The organization of this International Workshop was the responsibility of Dr. Victor N. Bushamuka, FAO Seed Security Consultant, and the preparation of the Workshop's proceedings was accomplished by Dr. William Fiebig, AGPS Agricultural Officer (Seed Security), both working under the technical guidance of Dr. U.G. Menini, Chief of the Seed and Plant Genetic Resources Service.

U.G. Menini
Chief, Seed and Plant Genetic Resources Service
Plant Production and Protection Division
FAO, Rome

LIST OF ACRONYMS

AGPS	Seed and Plant Genetic Resources Service (FAO)
CGR	Crop Genetic Resources
CBD	Convention on Biological Diversity
CGIAR	Consultative Group for International Agricultural Research
CIAT	International Centre for Tropical Agriculture
CIMMYT	International Maize and Wheat Improvement Centre
CIP	International Potato Centre
FAO	Food and Agriculture Organization of the United Nations
FEWS	Famine Early Warning System (USAID)
GIEWS	Geographic Information and Early Warning System (FAO)
GIS	Geographic Information System
GPA	Global Plan of Action
IARC	International Agricultural Research Center
ICRAF	International Centre for Research in Agroforestry
ICRC	International Committee of the Red Cross
ICRISAT	International Crops Research Institute for the Semi-Arid Tropics
IDP	Internally Displaced Persons
IDRC	International Development Research Centre (Canada)
IGO	Inter-Governmental Organization
INGO	International Non-Governmental Organization
IITA	International Institute for Tropical Agriculture
IPGRI	International Plant Genetic Resources Institute
IPR	Intellectual Property Rights
IRRI	International Rice Research Institute
ISTA	International Seed Testing Association
NARS	National Agricultural Research System
NGO	Non-Governmental Organization
OCHA	Office of Coordination of Humanitarian Affaires
OFDA	Office of Foreign Disaster Assistance (USA)
PGR	Plant Genetic Resources
PGRFA	Plant Genetic Resources for Food and Agriculture
PRAPACE	Programme Régional d'Amélioration de la Culture de la Pomme de Terre et de la Potate Douce en Afrique Centrale et de l'Est
PVO	Private Voluntary Organization
PVP	Plant Variety Protection
RESAPAC	Réseau pour l'Amélioration du Haricot (*Phaseolus vulgaris*) dans la Région de l'Afrique Centrale
SADC	Southern African Development Community
SINGER	System-wide Information Network for Genetic Resources
SOH	Seeds of Hope
SSCG	Seed Security Consultative Group
TCOR	Special Relief Operations Service (FAO)
UNDP	United Nations Development Programme
UNHCR	United Nations High Commissioner for Refugees
UNICEF	United Nations Children Fund
VAM	Vulnerability Analysis and Mapping
WANA	West Asia and North Africa (Network)
WARDA	West Africa Rice Development Association
WFP	World Food Programme of the United Nations
WIEWS	World Information and Early Warning System (FAO)
WVI	World Vision International

INTERNATIONAL WORKSHOP ON DEVELOPING INSTITUTIONAL AGREEMENTS AND CAPACITY TO ASSIST FARMERS IN DISASTER SITUATIONS TO RESTORE AGRICULTURAL SYSTEMS AND SEED SECURITY ACTIVITIES
(Project GCP/INT/660/NOR)
Rome, 3-5 November, 1998

PURPOSE, CONCLUSIONS AND RECOMMENDATIONS

1. INTRODUCTION

The incidence of man-made and natural disasters has increased in the past decades and had profound effects on rural communities and agricultural production systems, including farmer seed systems. Seed security is defined here as the access by farmers to adequate quantities of good quality seed of locally adapted varieties at the right time. Seed security is a prerequisite for the restoration of agricultural systems in disaster areas and a priority to ensure food security where people have access at all times to enough food to maintain an active and healthy life.

The conservation and sustainable use of plant genetic diversity is of capital importance and is currently the subject of many crucial debates. It must be stressed that crop genetic resources (CGR), which includes locally adapted and improved varieties, is especially endangered in disaster-prone areas and, in some cases, is the only source of CGR for a reasonably quick recovery and development of agricultural productivity by rural communities after disaster. Therefore, in disaster-prone countries and regions, the development and sustainability of agricultural production systems depends on the conservation of crop genetic diversity.

Farmers' access to seeds of locally adapted crop genetic resources is the foundation of present and future agricultural systems in developing countries. Therefore, seed security is a basic component of food security and is recognised as a priority by FAO and many other stakeholders, especially those in disaster-prone regions. This workshop was organised under FAO's mandate to facilitate the implementation of the Global Plan of Action (GPA) for the conservation and sustainable utilisation of crop genetic resources for food and agriculture, adopted by the International Technical Conference on Plant Genetic Resources, held in Leipzig, Germany in June, 1996. This workshop was also the follow-up of the International Workshop on Seed Security for Food Security, held in Florence, Italy, in November-December, 1997, where the importance of developing mechanisms to facilitate the restoration of local crop varieties after disasters was stressed as being critical to the re-establishment of seed supply and agricultural systems.

In this regard, the project GCP/INT/660/NOR, which is supported by the government of Norway, is designed to address the above issues. The mandate of this project includes the organisation of the international workshop during which strategies to assist farmers in disaster-prone countries will be developed.

2. PURPOSE AND OBJECTIVES OF THE WORKSHOP

The purpose of the workshop was to develop strategies and mechanisms that would strengthen seed security in developing countries and regions in disaster-prone areas, thereby enabling a rapid and effective restoration of farmer seed systems to restore domestic food security following disaster. The workshop brought together over 60 participants from the donor community, UN Organisations, CGIAR Centres, policy makers from disaster-prone countries, international organisations such as the ICRC, NGOs and seed relief specialists.

The objectives of the workshop were :

2.1. To develop strategies to improve the capacity of disaster-prone countries to better prepare farmer seed systems to minimise the effects of disasters. Through this workshop, mechanisms which enable countries and their rural farming communities to identify, acquire, multiply, and deliver seed of locally adapted varieties before, during, and after disasters will be identified.

2.2. To determine mechanisms of collaboration for a continuing partnership among governments and institutions involved in the maintenance or restoration of farmer seed systems, plant genetic resources, and seed security after disaster.

Workshop participants were instructed to recommend necessary follow-up actions requiring the collaboration of development agencies, plant genetic resources (PGR) institutions, emergency seed supply programmes, and the national and regional seed programmes to strengthen farmer seed systems by protecting local crop varieties from disasters as a critical component to restore agricultural systems and seed security.

3. GENERAL DISCUSSION

The workshop included a plenary session with the presentation of case studies and thematic working papers, two working group sessions and a final plenary session for discussion. Each of the presentations given in the initial session was discussed (see section 7) and provided a framework for in-depth discussion of issues in the working groups (see section 8).

Case studies that were presented were:

1. Angola / World Vision International;
2. Afghanistan;
3. Bosnia;
4. Rwanda / Seeds of Hope;

Background papers representing thematic issues were:

1. Characterising disasters;
2. Technical and socio-economic aspects of farmer seed systems;
3. Plant genetic resources and seed relief;
4. Regulatory aspects of seed security;
5. Seed stocks and seed multiplication in emergency situations;
6. Food and seed assistance in the recovery from crisis; and a
7. Synthesis of the background papers.

3.1. General comments

This international workshop marks an important step in approaching the sensitive problem of seed security. Perhaps for the first time, specialists from different regions of the world, in accepting FAO's invitation, are offering the benefit of their practical experience to help poor rural communities that are in need of a rapid response mechanism to save their local adapted crop varieties and to reproduce them in emergencies. This workshop intended, in fact, to provide a forum for exploring possibilities for defining seed security policies and for elaborating strategy guidelines for prompt intervention to be addressed particularly to farming communities in disaster-prone regions.

A seed security programme can be defined as a series of activities developed to ensure access of farming households - men and women - to adequate quantities of quality seeds and plant materials of adapted crop varieties at all times. In this definition, **access** implies that the source of these seeds should be within an acceptable distance, and supplies delivered in a timely manner and at affordable prices, while **at all times** refers to the availability of the appropriate seed stocks in time for each and every growing season and in rapid response to natural or man-made calamities.

Most farmers in developing countries are still largely self-provisioning in terms of local seed and plant material. Therefore, a seed security strategy should be conceived first in terms of **protecting the local crop diversity**, which is the foundation of present and future agricultural systems. However, the quality of these local seeds often needs also to be improved, and multiplication systems and distribution mechanisms upgraded to achieve food security. Thus, the second priority to guarantee seed security for food security can only be achieved if strategies and mechanisms are designed **to improve seed quality and supply systems**.

Taking action to restore seed security quickly after a disaster is an effective way to help restore national or regional food security. It would be even better if a seed security action plan could be implemented in anticipation of disaster. Therefore, the third priority in conceiving a seed security strategy could be **the development of an appropriate forecasting mechanism and an early warning system**. An alarm system, to forewarn about imminent breach of seed security, will provide an instrument for determining whether a seed rescue operation is needed in a particular community or country. However, it must be underlined that early warning systems require technical knowledge, intelligence-gathering, and effective collaboration among people, organizations, and nations.

Co-operative inter-country programmes and networks to foster a strong and continuing partnership among all the players concerned with maintaining seed security and/or restoring it after disaster as one of the approachable strategies were discussed in the workshop. A co-ordinating mechanism, such as a network, including national agricultural research systems (NARS) and other relevant national government agencies, regional organizations, non-government organizations (NGOs), inter-governmental organizations, (IGOs), private volunteer organizations (PVOs), International Agricultural Research Centers (IARCs), the private and public seed sector, farmers' organizations, communities and their respective associations, and international donors, could be a useful instrument to guide emergency programmes for the rehabilitation of agriculture in regions affected by calamities.

An effective seed security strategy comprises a series of mechanisms and activities that are supported by government policies to achieve seed security in a country or region. During the implementation of a seed security programme, it must be recognized that the primary determinants of seed security to farmers, their family and community are economic, cultural and ecological. To people affected by disaster, seed availability, accessibility and timely delivery are the overriding features to restore seed security, and consequently food security, at the household, community, national and regional levels. Thus, any strategy to foster the development of seed security schemes must fit the pragmatic aspects of the strategy to the socio-political framework that exists in the country or region.

3.2 Protection and conservation of crop genetic diversity

Protection and conservation of crop genetic diversity entails action to conserve locally adapted varieties and genetic resources at the farm and local community level, as well as in national and regional gene banks, and expanding stocks of the main varieties of regional food crops to ensure rapid seed multiplication and exchange during disasters. Also important are early warning systems to monitor changes in the status of genetic diversity in locally and regionally adapted crop genetic resources.

The common national and international response to calamities has been in form of emergency relief including seed supply for planting to restore agricultural systems in the affected areas. Although seed emergency relief programmes have helped millions of farmers in different developing countries, these programmes have been, in most cases, unable to restore the adapted crop diversity when lost in the region, and often introduce new varieties or only a few of the pre-disasters varieties. In addition to the risk of bringing new diseases and pests, introducing untested material into a region often results in yield reductions. The negative effects of inappropriate seed introduction may be long-lasting since most farmers in the developing countries traditionally save seed from previous harvests for the following cropping season. More importantly, untested material may pollute germplasm of the remaining local varieties, thus accelerating genetic erosion.

It is clear that emergency seed supply programmes require a substantial amount of technical and management planning to achieve the objective of restoring agricultural systems in regions victim of disaster. Therefore, a mechanism is needed to protect local crop varieties from disasters. The protected crop varieties could then be used as principle sources of seed and planting material for the emergency relief programmes. This would allow emergency relief agencies to distribute varieties that are needed by farmers and minimize technical mistakes for the benefit of both providers and recipients.

3.3 Seed supply systems

Robust seed supply systems depend on national and regional seed policies plus government and international support. Practical measures to strengthen seed supply systems include empowering on-farm and community seed production after disaster, farmers and their communities are frequently the lynch-pins in re-establishing local seed supply and seed distribution systems. There is also need for appropriate seed multiplication technology, including low cost, rapid biotechnology-based multiplication methods, strategic cost-effective seed reserves of the important food crops, and "sensible standards" to ensure production of high quality seed.

Institutions dealing with emergency situations must have the capacity to provide an efficient flow of information, an effective coordination capability and a state of readiness with appropriate mechanisms planned in advance. These types of institutions should have the following characteristics:

(i) a capacity to assess farmers needs before, during and after disaster in term of varieties and crops appropriate to the agro-ecological growing conditions of the affected region;

(ii) a capacity to collect seed stocks of appropriate varieties according to the potential magnitude and severity of the disaster, and to maintain these seed stocks when necessary; and

(iii) a capacity to multiply and distribute seeds and planting materials of adapted crop varieties that are needed by farmers in an equitable manner.

A number of UN agencies, international institutions (such as centres of the Consultative Group on International Agricultural Research), and NGOs, have recently been involved in restoring locally adapted seed and planting materials following emergencies. Unfortunately, these efforts have been, in many cases, completely ad hoc and voluntary in nature. There has been no standing capacity to respond appropriately to such disasters; no clear delineation of responsibilities; no coordinating mechanism to bring the various agencies and organizations together for planning and implementation of emergency seed supply efforts during and following disasters. Moreover, in some cases, agencies that supply seeds and planting materials as part of a relief package do so with little awareness or appreciation of the problems which can be caused by supplying inappropriate varieties.

These difficulties were recognized at the 1996 Inter-Governmental Leipzig Conference, which delineated the responsibilities of different agencies at the international level. The Global Plan of Action adopted at this conference gave the FAO the mandate to coordinate administratively a programmeme in this field in cooperation with WFP, UNHCR, UNDRO, IPGRI, national and international agricultural research centers, regional plant genetic resources networks, governments of the countries affected, donor countries and NGOs. In this endeavor, the FAO is specifically charged with establishing agreements with such partners for the "rapid acquisition and multiplication, restoration and provision of (planting) materials to countries in need."

It is therefore essential to elaborate this concept into practical schemes which will effectively address the needs of rural farming communities of member countries in regions affected by disasters.

3.4 Harmonization of seed rules and regulations

Sound national and regional seed policies are essential. Among measures discussed during the workshop are national plant improvement programmes with focus on locally adapted varieties and land races, a system to facilitate crop variety evaluation, registration and release, and informal, on-farm seed production and rural community distribution systems. At the regional level, policies should facilitate movement and exchange of seed across borders. This calls for regional seed security networks with databases on varieties grown within the region, harmonization of seed standards and regulations, and uniform regulations to encourage fair and equitable trade.

3.5 Seed Security Consultative Group

Finally, the workshop called for the establishment of a global Seed Security Consultative Group (SSCG) which, constituted under the aegis of FAO, would provide policy and technical advice for seed security, particularly in disaster-prone countries and regions.

This Seed Security Consultative Group (SSCG) is addressed to all regions and Member countries which have disaster-prone areas. The SSCG is open to all stakeholders involved in activities related to preparedness before, response during and rehabilitation of seed systems affected by disaster.

The SSCG will function on a voluntary basis and it will consist of networks of governmental and non-governmental organizations, international organizations, public and private organizations involved in the seed industry, international and national research centers, donors and other stakeholders involved in disaster response programmes in these regions.

Based on early warning systems information, the SSCG will serve to establish linkages with the above-mentioned field partners to address the necessary activities which have been identified to strengthen farmer seed systems which are faced with disaster, to respond during disasters and to rehabilitate affected seed systems in a sustainable manner to re-establish agricultural production systems. The field structure of the SSCG partners in any given disaster-prone region may vary. It may consist of existing national, regional, inter-regional networks involved with the seed sector, disaster relief and rehabilitation (see chart below).

The role of the SSCG will be to liase with its field partners in the region(s) affected by disasters in order to identify needed actions and assess comparative advantages in order to discern which partners will take which action. For example, who will take the lead in the identification of locally adapted crop genetic resources (CGR) and assure its conservation? If needed, who would be responsible for multiplication CGR needed to be distributed to disaster-affected farmers who need new seed and/or planting materials? Who would manage these distribution systems?

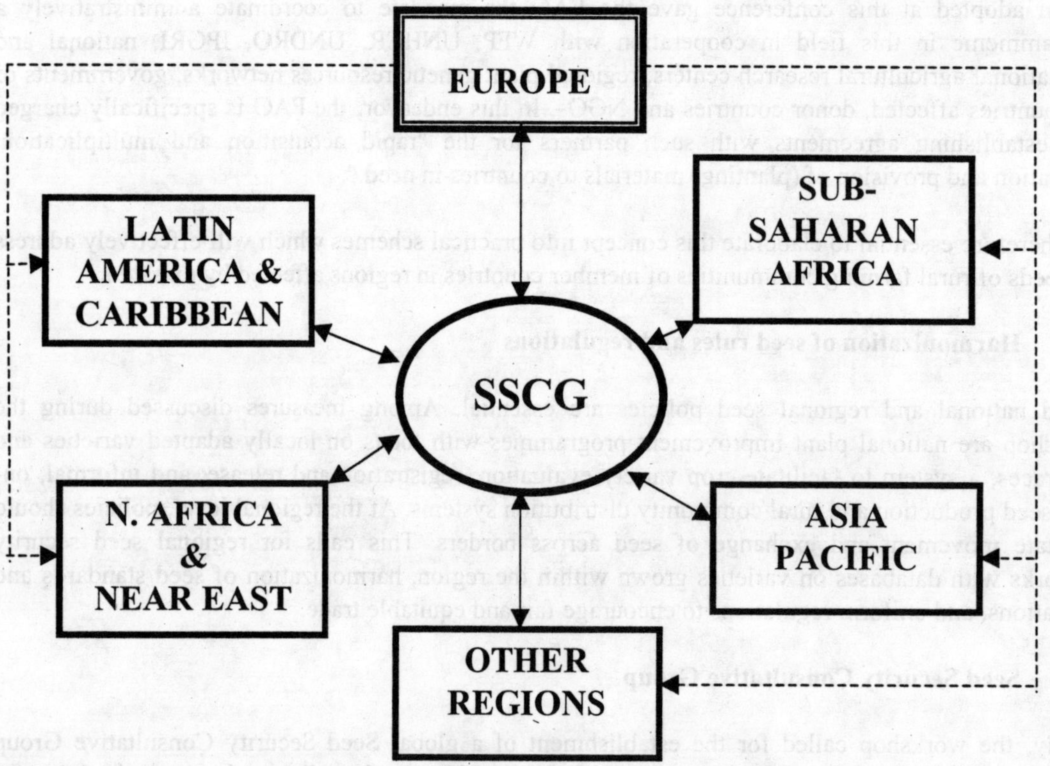

The type of coordination structure that is used for regional seed security programmes has much to do with how effective these programmes might be. It is ill advised to look for a group of high level administrators, such as government employees and other such professionals. The closer those responsible to coordinate seed security activities are to those who are most affected by food security problems, the more effective the leadership will most likely be.

At both the regional and national levels, it is much better to form a cooperative network of responsible people in organizations who want these programmes to work and to be effective in ensuring that farmers, men and women, have access to good quality seed and planting material for their preferred food crops in sufficient quantities, in a timely manner, and at an affordable price. These regional networks should be guided by a Coordinating Board which would be chaired by a Regional Coordinator and include representatives of the various field partners. FAO is well suited to assist member countries in designing appropriate advisory mechanisms for their seed security programmes in disaster-prone regions.

4. CONCLUSIONS OF THE WORKSHOP

The workshop concluded that:

- International efforts to support seed system recovery have sometimes been inconsistent.
- Information on CGR, if available, is often not accessible to stakeholders confronted with emergency situations.
- Impacts of disasters on farmer seed systems are often complex and varied.

- Experience has shown that important CGR can be lost in the wake of a disaster.
- Responses to disaster require collaborative and coherent approaches to restore seed systems.
- Laws affecting seed movements can facilitate or constrain the emergency response efforts to restore seed systems.
- Inappropriate emergency response efforts to restore seed systems can result in further disruption of seed systems.
- Inadequate disaster response has often resulted from lack of preparedness.

5. CHALLENGES AND RECOMMENDATIONS OF THE WORKSHOP

The following specific actions were recommended:

5.1 Seed Security Consultative Group

- FAO should seriously explore the possibility to support the establishment of a mechanism which could take the form of a Seed Security Consultative Group, under the auspices of FAO, with a broad constituency including civil society, research and extension institutions, NGOs, farmers groups and public/private seed sectors.
- This mechanism should optimise the comparative advantages of the various stakeholders to assure seed security in disaster-prone regions.

5.2 Information

- Inventories of CGR should be gathered and consolidated.
- Existing electronic information management systems on CGR should be made more user friendly and accessible to stakeholders.
- Capacity of the stakeholders to use these information systems should be strengthened.

5.3 Seed security

- Explore methodologies which assure effective seed security needs assessment by stakeholders involved in emergency response.
- Support gender sensitive diagnosis of seed systems (formal as well farmer based) existing prior to the disaster.
- Establish guidelines and checklists in order to facilitate coherence and comprehensiveness of assessments.
- Ensure that seed system assessment is linked to other sectorial assessments, including food security, etc.

5.4 Crop genetic resources (CGR)

- *Ex situ* genebanks have an essential role to play in preserving CGR and resources for the effective functioning should be available to them.

- CGR stocks should be securely conserved and made available in appropriate quantities, if needed.

- Seed system recovery programmes should make the provision to restore farmer's CGR affected by disaster.

5.5 Seed networks

- A mechanism should be promoted to co-ordinate procurement, multiplication and distribution of quality seed of locally adapted varieties at the national and/or regional level. FAO can play an important role here.

- As mentioned in the GPA, activities preventing the breakdown and supporting the re-emergence of all seed supply networks are to be given a high priority.

5.6 Rules and regulations

- FAO should explore the possibility to establish a mechanism that will encourage and co-ordinate regional harmonisation of seed rules and regulations. FAO will promote the ideas that all stakeholders be involved in that process.

- Proper attention should be paid to existing UN sanction mechanisms so that seed provision is not hampered.

5.7 Learn from past experience for appropriate response strategies to emergencies

- FAO should encourage the assessment of impacts of interventions and dissemination of lessons learned.

- Case studies should be carried out and their results (lessons learned) shared.

5.8 Preparedness

- Disaster prevention and preparedness measures should receive proper attention from all stakeholders, especially donors.

6. SYNOPSIS OF CONCLUSIONS AND RECOMMENDATIONS OF THE WORKSHOP

CONCLUSIONS	RECOMMENDATIONS
Conclusion 1. International efforts to support seed system recovery has sometimes been inconsistent.	**Recommendation 1.1.** FAO should seriously explore the possibility to support the establishment of a mechanism which could take the form of a Seed Security Consultative Group, under the auspices of FAO, with a broad constituency including civil society, research institutions, breeders, farmers groups. **Recommendation 1.2.** This mechanism should optimise the comparative advantages of the various stakeholders to assure seed security.
Conclusion 2. Information on CGR is often not accessible to stakeholders confronted with emergency situations.	**Recommendation 2.1.** Inventories of CGR should be continued and consolidated. **Recommendation 2.2.** Existing information on CGR should be made more user friendly and accessible to stakeholders. **Recommendation 2.3.** Capacity of the stakeholders to use these information systems should be strengthened .
Conclusion 3. Impacts of disasters on farmer seed systems are often complex and varied.	**Recommendation 3.1.** Explore methodologies which assure effective seed security needs assessment by stakeholders involved in emergency response. **Recommendation 3.2.** Support gender sensitive diagnosis of seed systems (formal as well farmer based) existing prior to the disaster. **Recommendation 3.3.** Establish guidelines and checklists in order to facilitate coherence and comprehensiveness of assessments. **Recommendation 3.4.** Ensure that seed system assessment is linked to other sectorial assessments, including food security, etc.
Conclusion 4. Experience has shown that important CGR can be lost in the wake of a disaster.	**Recommendation 4.1.** *Ex situ* genebanks have an essential role to play in preserving CGR and resources for the effective functioning should be available to them. **Recommendation 4.2.** CGR stocks should be securely conserved and made available in appropriate quantities, if needed. **Recommendation 4.3.** Seed system recovery programmes should make the provision to restore CGR affected by disaster.
Conclusion 5. Responses to disaster require collaborative and coherent approaches to restore seed systems.	**Recommendation 5.1.** A mechanism should be promoted to co-ordinate procurement, multiplication and distribution of quality seed of locally adapted varieties at the national and/or regional level. FAO can play an important role here. **Recommendation 5.2.** As mentioned in the GPA, activities preventing the breakdown and supporting the re-emergence of all seed supply networks are to be given a high priority.
Conclusion 6. All laws affecting seed movements can facilitate or constrain the emergency response efforts to restore seed systems.	**Recommendation 6.1.** FAO should explore the possibility to establish a mechanism that will encourage and co-ordinate regional harmonisation of seed rules and regulations. FAO will promote the ideas that all stakeholders be involved in that process. **Recommendation 6.2.** Proper attention should be paid to existing UN sanction mechanisms so that seed provision is not hampered.
Conclusion 7. Inappropriate emergency response efforts to restore seed systems can result in further disruption of seed systems.	**Recommendation 7.1.** FAO should encourage the assessment of impacts of interventions and dissemination of lessons learned. **Recommendation 7.2.** Case studies should be carried out and their results (lessons learned) shared.
Conclusion 8. Inadequate disaster response has often resulted from lack of preparedness.	**Recommendation 8.** Disaster prevention and preparedness measures should receive proper attention from all stakeholders, especially donors.

7. SUMMARY OF WORKSHOP PLENARY DEBATE

7.1. Presentation of case studies

Five case studies were presented based on field experiences in seed relief and seed system rehabilitation in Angola, Afghanistan, Bosnia, with two case studies from Rwanda by FAO/TCOR and the Seeds of Hope project. Based on the analyses of the authors and the discussions of the studies, a consistent set of issues emerged as lessons learned from their experiences. The description of the case studies and some of the main lessons learned are described below.

7.1.1. Angola

World Vision International (WVI) has been working in Angola in the agricultural sector for half a decade. The operation has gone through a series of phases. The first one was based on a standard seed kit involving maize, sorghum, beans and vegetable seeds. Seeds were imported from neighbouring countries. World Vision International, as many other agencies involved in seed and tool distribution programmes, realised the difficulties to work with a standard kit of imported seeds in a highly ecologically diversified country. This led to a series of evolutions of the programmeme, which encompassed on-station experiments to screen varieties on the basis of their characteristics *vis-à-vis* of drought tolerance, resistance to pests and diseases, etc. Important criteria were also cooking and taste-related characteristics. As part of the challenging "Seed of Freedom project", WVI entered into a collaborative effort with International Research Centres in order to identify adapted varieties to be included in the distribution list and to retrieve some of the varieties which have been lost in the wake of the three decades of civil war. World Vision International also realised the importance of roots and tubers, often forgotten in the distribution of agricultural inputs. But after the multiplication of this genetic material, the efficiency of the distribution/dissemination mechanisms emerged as a pressing issue. Support to farmer co-operatives, involvement of farmers in the whole research process and capacity building of extension workers and researchers were seen as the most effective tools for these endeavours.

Lessons learned:

- **Identification and availability of appropriate seed stocks of locally adapted varieties is essential to restoring seed systems.** World Vision International in Angola found that the success of helping rural farming communities achieve a certain level of food security depended significantly on which varieties were included in their seed distribution programmeme. In many cases, seeds of inappropriate varieties had the effect of retarding the pace of agricultural recovery because many of the varieties performed poorly in Angolan environments for which they were not adapted.
- **The distribution of grain as seeds jeopardises the recovery.** The use of grain stocks as seeds can have a negative impact on production by introducing varieties which are not adapted and also lack of quality control.
- **A preliminary assessment identified that the primary need of displaced and returning farmers was for animal traction and hand tools, not necessarily seeds.** They needed help to cultivate larger areas and a sufficient number of tools to allow several members of a family to work in the field simultaneously, while seed stocks of major food crops usually could be obtained locally, one way or another.

7.1.2. Afghanistan

The state of the agricultural sector of Afghanistan after two decades of war is diversified and complex. On the one hand, many areas are reaching near subsistence level. On the other, Afghanistan, which used to export large quantities of grains, is currently receiving important tonnage of food aid.

Some areas are still heavily mined, while others have been diverted from cereals to poppy growing. Yet, there is no image of dramatic famine and the Afghan people have maintained their tradition of hospitality. After the 1988 peace agreement, the early years were marked with donor generosity. This faded away with the civil war continuing. The FAO agricultural programmes have managed to continue albeit on a reduced scale. A series of constraints had to be overcome, such as ongoing fighting preventing access and political games played by local commanders.

Lessons learned:

- **In the emergency relief programme, crop seeds and fertilisers were as important to farmers as other goods, such as food and medicines.**
- **The way aid is provided should be chosen on a case by case basis: for instance, when the State is disrupted by the war and not functioning, the recovery is possible only with direct support to farmers/communities.**
- **The focus on specific areas and problems related to seed production and distribution then allowed the support to be effective.**
- **The project was successful because of the holistic approach used, since the farmers do not differentiate among the different phases of emergency relief, rehabilitation and development. The Afghan experience underlined the importance to have the support to the recovery of the food production sector included into the "emergency operation", rather than in the "rehabilitation agenda". It has to be perceived as a support to the survival strategy in the midst of the crisis, rather than a set of activities for the more quiet times.**
- **Need for a flexible seed development programmeme.** Because of the complexity of the socio-political environment, a flexible seed programmeme of diversification, sustainability and collaboration with other UN agencies was necessary.

7.1.3. Bosnia

Different from the African or Asian systems where farmers normally rely mainly on on-farm saved seeds, agricultural systems in Bosnia prior to the war depended on seeds produced in the main seed production areas of Voivodina, in the rich Danau-Save plain. When the war broke out, these seed supply channels were made dysfunctional by the frontlines and later by the establishment of the new Frontiers of Serbia and Bosnia-Herzegovina. As early as 1994, FAO, UNHCR, ICRC and NGOs have been involved in seed distributions of various types in order to reinforce resilience both at the family level and at the co-operative level. As these programmes were evaluated, two main issues came up: 1) Bosnia was ill equipped in terms of laboratories and equipment; and 2) the fact that the country was so heavily dependent on seed import was seen as implying continued vulnerability. FAO, alongside with its seed distribution programmes, embarked in two challenging directions: to address the quality issue and to tackle the sustainability question.

Lessons learned:

- **Disasters do not affect formal and informal seed systems in the same manner. Efforts to rehabilitate these systems differ in approach and consideration of these system's needs should be defined as soon as the situation allows in disaster-affected areas.** Farmers depended almost entirely on the formal seed sector for their provision of seeds. During and after the war, this supply system was devastated and FAO had to focus on the rehabilitation of seed multiplication and quality control capabilities.
- **Co-ordination issues were paramount.** The sophistication of such a type of programmeme also implies that the human resource factor be approached in an appropriate way. Co-ordination with other agencies prevents overlapping of efforts. The building up of national capabilities has to receive special attention.

- **Rapidity of action allowed to accomplish the tasks.** Quick response from FAO for approval or rejection of requests from the field allowed rapid decision making and action.
- **Continued support and follow-up activities are necessary.** It was noted that unfortunately, when the emergency phase is nearing end, it is much more difficult to attract donors attention for sustained support of the still fragile emerging formal seed sector.

7.1.4. Rwanda

After the civil war that affected Rwanda in early 1994, impressive efforts were made to ensure that seeds and tools were distributed to farmers. In order to limit duplication and prevent technical errors, FAO together with the newly established Ministry of Agriculture, set-up a co-ordination mechanism where UN agencies, government institutions, ICRC, NGOs and donors could meet and co-ordinate their activities. FAO's knowledge of the area was crucial in facilitating the elaboration of a set of technical recommendations for seed distribution. Indeed, in view of the complexity of Rwanda's agro-ecosystem, many agencies did not know how to find seeds well suited to farmers needs. It was, for instance, important to recommend the distribution of mixtures of bean seeds rather than pure cultivars. Timeliness of the distributions to fit within the crop calendars and comprehensiveness of the coverage in order to limit the creation of pockets of food insecurity were two crucial outputs of this co-ordination. FAO started a quality declared seed multiplication programmeme in order to restore the national capacity and pave the way toward more sustainability in the agricultural sector.

An interesting and important feature of the international emergency response for Rwanda is the creation of the "Seeds of Hope" project (CIAT) which based its activities on two main lines: re-introduction of varieties and land-races adapted to the agro-ecological conditions of Rwanda and to provide technical assistance to rebuild seed and crop production capacity in Rwanda. A thorough survey was initially carried out in order to get an up-to-date understanding of what was really the situation of the seed systems in Rwanda and how national and international agricultural research centres could help restore the pre-crisis diversity and local varieties. "Seeds of Hope", as a knowledge based enterprise, permitted significant methodological improvement in the overall approach to rehabilitating seed systems affected by the disaster.

The field research underlines the importance to recognise the existence of both the formal and the informal seed systems. The first one is based on the activities of specialised breeders and seed producers supported by extension and research institutions. The second is farmer-based and largely relies on social networks and small scale trading and bartering mechanisms. The research carried out by "Seeds of Hope" underlined the difference between "seed security" and "varietal security".

Lessons learned:

- **The informal sector should receive adequate support**. In Rwanda, 90 per cent of the seeds that farmers use comes from the informal sector. Therefore, this important sector should receive proper attention.
- **Disasters do not affect formal and informal seed systems in the same manner, nor do they play the same role in farming system recovery.** Efforts to rehabilitate these systems should be appropriate to the specific needs as early as possible, and should be identified during diagnosis phase linked to need assessment. The "Seeds of Hope" project's reintroduction of local land-races and improved varieties adapted to the agro-ecological conditions of Rwanda and technical assistance to rebuild seed and crop production capabilities led to an awareness of the difference between "seed security" and "variety security". Some farmers had lost their seed stocks of their preferred local varieties, but seeds were still available of these varieties in the country. In other cases, local varieties of certain crops were totally lost.
- **Effective co-ordination of stakeholder's efforts (donor, UN Agencies, governmental and non-governmental) is critical to rehabilitate seed systems.** FAO's experience in Rwanda

clearly showed that a mechanism to co-ordinate UN Agencies, government institutions, NGOs, and other stakeholders in disaster relief activities to restore formal and informal seed systems resulted in effective and efficient interventions in procuring, multiplying and distributing seed stocks of adapted varieties to disaster-affected communities.

- **A diagnosis and needs assessment of the impact of disasters on seed systems should be completed before seed relief programmes are designed and implemented.** In order to avoid an all too common mistake made in emergency responses of purchasing and distributing seed of inappropriate varieties, it is essential to assess the impact of the disaster on affected rural communities. Farmer seed systems can be affected in many ways and not all crops are affected in the same manner.
- **Seed rules and regulations can facilitate or hinder the seed procurement process.** It is important that sound rules and regulations are established. Over-prescriptive regulations can hinder the seed procurement process. On the other hand, the lack of a national policy and effective seed rules and regulations hinder the efforts made by the public and private sectors to re-establish seed supply channels.

7.2. Presentation of thematic working papers

Six background papers were prepared which addressed the main issues related to disaster situations which affect seed systems, actions needed to provide farmers assistance under these circumstances and efforts required to restore agricultural systems and seed security activities. The following sections were highlight the main aspects to consider in emergency relief and recovery programmes.

7.2.1. Characterising disasters

- **It is important to analyse the time characteristics of disasters.** Indeed, it is important to know not only if a disaster is man-made or a natural one, but also to analyse its time characteristics: if it is rapid or slow, frequent or rare, of short or long duration.
- **It is important to consider the place where the disaster occurs.** The same disaster would have a different impact whether it occurs in a developed country, where farmers are normally buying their seeds from seed companies, or in a landlocked developing area where the prevalent feature is farmers reliance on their own saved seed.
- **The assessment phase is crucial for the success of the relief operations.** It appeared that the quality of the diagnosis of the situation of the farming systems and seed systems prior to the crisis, of the analysis of the disaster itself and of its impact are key factors for appropriate intervention design, planning and implementation.
- **Early warning systems are paramount, but need to be complemented by effective preparedness and followed by prompt action.** The relevance of early warning systems was considered. Experience shows that, in fact, most of the recent disasters were not only predictable, but also predicted. The missing elements were either effective pre-disaster preparedness, or lack of political will to intervene. It is preferable to try to mitigate the effect of a disaster by taking action before it happens.

7.2.2. Farmer seed systems in disaster situations

- **A holistic approach should be used in dealing with farmer seed systems.** This subject was considered from two complementary perspectives: the technical aspects and the socio-economic angle. One assumption is often forgotten: farmer seed systems are, in most instances, part of the farmer's over-all production system, with part of the harvest saved for future planting. Thus, the impact of disasters on farmer seed systems cannot be analysed outside of the entire farming system.
- **Farmer seed systems are a key point of relief programmes.** Since crop and variety diversity play a crucial role in risk minimisation and disaster mitigation, the support to the farmer seed

system shall be seen as central to survival mechanisms, recovery dynamic and prevention of recurrence of a disaster.

- **Farmers' access to preferred varieties should be guaranteed.** Varieties matching the actual local needs can be identified only according to farmers' preferences, which are often based on characteristics differing from the technical criteria used by seed specialists.
- **Farmer seed systems are not only very diversified, but they also change over time.** In an emergency period, these changes can even constitute important ways in which farmers respond to a crisis. Farmer seed systems are very resilient and farmer coping strategies and innovative capacity often surprise the "emergency seed suppliers".
- **Good co-ordination of stakeholders is necessary.** Many stakeholders are involved in supporting the seed sector in disaster and post disaster situations and without co-ordination, their efforts may not effectively achieve desired goals. Institutional collaboration is deemed necessary.

7.2.3. Plant genetic resources and seed relief

- **Put more emphasis on the resilience of farmer seed systems and on the protection of intellectual property rights (IPR).** The interest of plant breeders towards crop genetic diversity is not new. It has nevertheless been more "breeder oriented", while the new focus of this specific domain of bio-diversity is more on supporting resilience of farming systems as well as protecting farmers' CGR.
- **Linkages are needed (at national, regional and international level) between those involved in provision of seed in disaster situations and those involved in management of genetic resources.**
- **Strengthen regional PGR networks and enhance their capacity to manage information on genetic resources availability in *ex situ* collections for seed relief programmes.**
- **Further investigations, through collaborative studies, are needed to determine how PGR information can best be used, what its strengths and weaknesses are in the context of the activities envisaged and what improved systems could be developed.**

7.2.4. Regulatory aspects of seed security

- **Seed rules and regulations should support seed quality control and the functioning of sound seed systems, not police them.** This relatively provocative statement served as a conclusion of a rather innovative and stimulating presentation on how seed laws can help or hinder relief and recovery activities in times of disasters and post-disaster.
- **Seed regulations which do not match with the actual national seed situation can hamper, or prevent, relief programmes.** If it is normal that countries try to protect themselves against the risks entailed in uncontrolled movements of genetic materials, it is also to be noted that sometimes, even the trade or distribution of seeds from local grain markets is prohibited since these local seeds are rarely registered, much the less certified.
- **The importance of phytosanitary measures and an early involvement of the phytosanitary service were stressed.** The importance of phytosanitary regulations was not questioned. There is little worse than an ill conceived and badly controlled relief operation resulting into the spreading of new diseases or unknown pest, or new weed.
- **Good communication with seed quality control officials avoids certification problems.**

7.2.5. Seed stocks and seed multiplication in emergency situations

- **Preparedness is the key for success.** In order to respond effectively and with cost-effective measures in emergency situations, it is necessary to be prepared and know: 1) what needs to be replaced; 2) where to get the appropriate seed stocks; 3) how to multiply; and 4) how to distribute the seeds.

- **Danger: the loss of genetic resources.** Erosion of stored plant genetic material depends on the quality of the original material and on the conditions under which the material is maintained and periodically multiplied. Loss of genetic resources may also occur in farmers' fields and in nature. Information management systems for CGR can provide very useful information about the preventive, or remedial actions, to be taken needed to restore these CGR affected by disasters.
- **Early warning system: a must.** An early warning system for CGR is a necessary component in a seed security strategy, to allow rapid action and response.
- **Information about crops matching climatic conditions is available.** There are also a number of tools to predict crop growth responses to existing climatic conditions. These preventive measures would assure that appropriate seed stocks are identified, procured and multiplied to restore seed systems affected by disasters.

7.2.6. Food and seed assistance in recovery

- **Crop seeds and food are strictly connected and should be provided simultaneously.** The loss of access to seeds and food are often interconnected. While seeds are crucial to agricultural recovery, human energy to plough the land is equally important. The international experience on post-disaster recovery shows that instead of looking at the intervention cycle with a food phase and a seed phase, it is often much more efficient to combine both resources as a means to rebuild rural economies and to launch agricultural activities in the aftermath of the disaster. Seed relief is thus viewed as integral part of the emergency package.
- **Farmers realise the vital importance to preserve crop seeds for agriculture.** Many programmes have demonstrated that, if the seeds distributed is not clearly labelled and distinguishable from the food ration. seeds of important land-races and expensive imported seed can be consumed alike, with the farmers often preferring to eat the distributed seeds and to keep their own stocks. Integrated seed and tool provision is therefore not only useful, but often essential for the successful resumption of agricultural cycles, including the rehabilitation of local seed systems.
- **Collaborative efforts between food relief agencies and actors of the agriculture sector should start as early as possible in the planning stage of the response to the disaster.**

8. RESULTS OF WORKING GROUP SESSIONS

Each of two working groups was assigned a topic: the first group had to deal with technical aspects and the second group with institutional aspects related to rehabilitating seed systems affected by disasters.

8.1. Technical Working Group

Topic: Mechanisms which enable countries and their rural farming communities to identify, acquire, multiply and deliver seeds of locally adapted varieties before, during and after disasters.

The group split the issues into five key working points:

Point 1: look at the possibilities to develop guidelines / checklists / simple explanation for donors/ relief agency policy makers on "what is variety-sensitive seed relief";
Point 2: analyse how seed regulation can be better adapted to disaster situations;
Point 3: identify the ways and means by which a "CGR" system could help stakeholders to cope with disasters;
Point 4: identify suitable co-ordination mechanisms to support efforts specifically geared to strengthen farmer seed systems. In that specific mechanism, what would be the role of FAO?; and

Point 5: how to improve capacity for disaster preparedness and response? What could be the modalities for funding these proposals?

In order to carry out its work and deliver conclusions and options for recommendations, the working group organised its work in four phases: 1) quick review of all issues; 2) work in smaller settings to be more operational; 3) regrouping and presentation of the work of the sub-working groups; and 4) debate and finalisation for presentation to the Plenary.

Point 1:

The working group thought that both dissemination of ideas through a "simple explanation paper for the donors" was very relevant in view of their frequent reluctance to look in a comprehensive manner to seed issues and to consider that beyond simple seed distribution programme, there is the much broader question of the recovery of the seed systems. The group also considered that the drafting of guidelines and checklists would be very useful in order to help stakeholders to avoid mistakes and to facilitate implementation of programmes. However, after an initial look at the time requirements for the elaboration and drafting of the three types of documents, the working group decided to focus on the guidelines.

In the discussion about what should be the main points covered in these guidelines, five main issues were outlined:

- Seed systems are encapsulated within social, economic and farming systems. The understanding of the context with all the different factors is paramount;
- This implies that quality diagnosis of the situation prior to the disaster, analysis of the disaster and of its impact on agriculture and food security in general and on seed systems in particular is the first task to carry out;
- This might come with various types of recommendations, including not to touch the seed system or to approach seed security from a non-seed angle (through food distribution for instance);
- If the response is "yes, we need to support the seed system", then a multi-prong strategy might have to be designed, including appropriate seed relief, support to farmers capacity, involvement of appropriate and competent PGR institutions; and
- There is a need to link short and longer term considerations for seed system rehabilitation and sustainable development in both the formal and informal seed sector.

On this basis, a draft text was prepared by a small sub-working group and discussed. After minor adjustments and precision, it was finalised for distribution at the plenary (see Table 1 below). The "Guidelines" might be further developed and finalised by the iterative exchange between FAO Secretariat (AGPS) and a network of experts. A smaller group could be given the task to prepare a text for the "small explanation for donors" which could be finalised following the same iterative process.

Point 2:

The seed regulatory framework, which may include variety release procedures, seed quality control, phytosanitary certification, International Property Rights (IPR) and access laws, is a complex set of laws and regulations. It should facilitate access to seeds of appropriate varieties and sufficient qualities by farmers in normal or abnormal conditions. This set of rules and regulations should support the development of both the informal and formal (public and private) seed systems in accordance to the national seed policy.

Table 1. Guidelines for Funding "Variety-Sensitive Seed Relief" Activities

1.	Demand sound analysis of the socio-economic/political/environmental context of the disaster. Include some analysis of the pre- and post- situation.
2.	Conduct a focused needs assessment (qualitative and quantitative) of the kinds of seed systems support which might be needed, if any. Be sure to include analysis of the capacity of informal (local) and formal seed systems in their changing context. This includes analysis of more than seed itself. It should include analysis of the ability of farmers to procure seed, plant and manage their crop, i.e. to carry on successful agricultural production.
3.	Choose varieties of major food crops, which are sufficiently adapted to the agro-ecological and socio-economic conditions of the disaster-affected area.
4.	Ensure that the implementing agency has access to technical expertise to procure, evaluate, store and deliver good quality seed in a timely manner.
5.	Strategies for seed systems rehabilitation should be carefully phased. While in the first phase, a focus should be on the provision of seed stocks of "adapted" and "acceptable" varieties may be sufficient, subsequent phases should plan for self-sustaining strategies by which farmers can produce and save/procure seeds of their diverse and preferred varieties.
6.	Each season after disaster relief, the capacity of the formal and informal seed supply channels should be re-assessed. In some cases, creating, rehabilitating or strengthening local seed supply channels may be sufficient.
7.	Efforts should be made to co-ordinate relief activities with complementary agencies and organizations to avoid duplication of efforts. If appropriate, lead agencies should be identified.
8.	Records of all varieties procured and seed distribution of those varieties should be made and carefully managed. Monitoring and evaluation must be programmed on seed performance and variety adaptation within the farming systems in which they are distributed. This quantitative and qualitative information should be geo-referenced and stored in relational databases at the national, regional and global levels.
9.	A transitional strategy to move from the acute emergency response to a more non-dependent and capacity building phase during the rehabilitation of formal and informal seed systems. These strategies should be components of intervention policy and programmes during the design phase and be based on lessons learned.

The working group recognised that these rules and regulations often have the tendency to hamper the effectiveness of support to seed systems in disaster situations. They often result in unnecessary hindrance to the development of diverse seed systems and to appropriate responses to emergencies. In the process of the regulatory reform, due attention should be given to the needs of the seed relief and seed systems rehabilitation during and after emergency.

The working group identified various points deserving attention:

- International harmonisation of rules and regulations, including their interpretation and implementation is important;

- In this respect, special reference should be made to phytosanitary regulations. Indeed in a disaster affected areas, there is no need to introduce further problems with new diseases, new weeds, etc.;
- Transparency of the process and representation of all stakeholders in the response to the disaster is important in order to ensure that these rules and regulations will be respected;
- Avoidance of unnecessary restrictions was seen as crucial, while strengthening seed quality awareness during procurement was seen as a very important tool toward assuring overall seed quality and appropriateness; and
- Any development of IPR and farmer rights should avoid hindrance to seed systems in times of emergency.

Then the group came with a few suggestions:

- The forthcoming Regional Seed Policy and Programmes meetings should address these recommendations of the Workshop in relation to seed rules and regulations; and
- FAO should play an important role in that process by recommending member countries to review their existing seed rules and regulations. A survey of existing systems is recommended to identify the nature and variety of the problems as well as to identify best practices and success stories.

Point 3:

A first part of the discussion focused on identifying what were the components of a CGR system.

The two main components of a CGR system in emergency and post-emergency situation could be, but not limited to:

1) <u>Information Management System</u>: It was recognised that in order to locate, and if necessary to replace lost germplasm, information on location, characteristics and agro-ecological and socio-economic profiles of the germplasm would be required. This knowledge would have to be based on inventories and characterisation of CGR grown in the area before the disaster. The meeting recognised that in order to do this, there is a need to develop simple passport data based on a minimum number of important descriptors. All this information could be accumulated within a user-friendly electronic information management system. Mechanisms for the information to be made accessible were deemed essential since the timeliness factor is very important in the context of disaster.

2) <u>Material</u>: Characteristics of possible sources for germplasm recovery for eventual re-multiplication, bulking and distribution of lost varieties were identified. Both CGR *ex situ* and *in situ* conservation can play a role in that matter, although the very small quantities of seeds kept in *ex situ* gene banks does not permit rapid multiplication and distribution.

Point 4:

The working group identified mechanisms to support efforts specifically geared to strengthen farmer seed systems

- Facilitate dialogue between countries on policy matters related to seed sector development, such as the upcoming FAO regional technical meetings on Seed Policy and Programmes;
- Continuing discussion of issues, development of conceptual framework, technical advice; and
- Providing administrative/operational co-ordination.

The working group came to the conclusion that the recommendation made at the Florence meeting for the establishment of a Seed Security Consultative Group was still relevant. One option would be a core group, to be formally established, and a wider, informal support network. Another option would be to establish a Seed Security Working Group with a very wide and acknowledged constituency where actors from the civil society, technical bodies, NGOs and international agencies would be represented. There would be a need to draw up terms of reference for the core group and network.

Point 5:

A four-point strategy was prepared to improve disaster preparedness and responses.

1. Increase participation of civil society (build links between technical agents, local NGOs, farmers groups, social movements concerned with land and seed rights, refugees and displaced groups);
2. Disaster victims should be more actively engaged, when and where feasible, in activities directed at securing their seed resources and assessing needs to rehabilitate their seed systems. Although there are often many constraints imposed on the access of refugees or IDPs to land, it has been seen that when displaced people come back home, the most difficult thing for them is to rebuild their own CGR and to recover their old varieties.
3. Make information from PGR and CGR data bases and collections more known and user friendly. They should indeed be much more accessible to a range of potential stakeholders. Potential users of this information should be trained.
4. Pay increase attention to risk minimisation in seed recovery and rehabilitating agriculture.

8.2. Institutional Working Group

Topic: Mechanisms to strengthen collaboration for a continuing partnership among governments and institutions involved in the maintenance, restoration of farmer seed systems, plant genetic resources and seed security after disaster.

The first consideration taken by the working group was the level at which the intervention would take place:

1) local;
2) community/district/province;
3) national;
4) regional; and
5) international/global.

For each of those levels the group looked at the stakeholders who would be involved. Some stakeholders go across all levels and others operate only at one, or more. This analysis is summarised in Table 2. That gives a two-dimensional matrix within which to consider the institutional structures that have to be in place. Analysing that matrix, the group identified actions necessary in emergency situations. These actions are performed at different stages: before, during and after the disaster (Table 3).

The most cost-effective effort stakeholder involved in disaster response activities can make, is to focus on preparedness. Identifying sources of seeds for varieties which are grown in disaster-prone areas is essential to identifying varieties which are adapted to the agro-ecological and socio-economic conditions of farming systems facing potential disasters. Making effective use of early warning information is critical to being prepared before the disaster occurs.

At the onset of disasters, a thorough diagnosis and needs assessment of the farming systems affected can identify which responses are most appropriate to mitigate the impacts of the disaster on existing seed systems. If procurement of seed stocks is necessary, knowledge about the existing mechanism for the effective multiplication (if needed) and distribution of these seed stocks to communities in need will lead to more effective responses.

Impact assessments of the efforts to rehabilitate disaster-affected seed systems will lead to the identification of strategies to assure that these seed systems are not left in such a fragile state that once relief efforts end, seed insecurity is still potentially a problem. These assessments should also be documented as lessons learned about what worked to assure seed security and why, but more importantly, what did not work well and why not.

As a consequence, the group came out with a number of recommendations to develop an institutional framework that can help address the issues of emergencies and disasters, particularly in respect of the seed systems (Table 4).

The first recommendation, is that FAO establishes a mechanism that will assemble and manage information on crop genetic resources from the stakeholders involved. This mechanism should facilitate the making of information on crop genetic resources and seed systems readily available to the stakeholders.

The second recommendation is that FAO explores the possibility to support seed security needs assessment by stakeholders involved in the emergency response. In many cases, stakeholders are on the ground when the emergency breaks, including country nationals, NGOs, FAO representatives and members of national international agricultural research centres.

The third recommendation is that FAO establishes a mechanism to encourage and co-ordinate regional harmonisation of seed rules and regulations. Nothing is more important, during an emergency, than to facilitate the sensible movement and distribution of seeds across national borders, if necessary.

The fourth recommendation is that a mechanism is establish to co-ordinate the procurement, the multiplication and the distribution of good quality seeds of adapted varieties at the national and regional levels.

The working group came to the collective decision, based on recent experiences, that the technological know-how to procure, multiply and distribute seeds exists. This knowledge must be disseminated to all stakeholders in a user-friendly format as best practices based on lessons learned.

Finally, the group concluded that as disasters are going to be a continuing problem, mechanisms must allow for continual build-up of information and knowledge, know-how and expertise to improve the speed, rapidity and effectiveness of the response to disaster. Thus, it is necessary to encourage the assessment of the impact of the interventions to restore seed systems after disasters.

Table 2. Stakeholder analysis at different socio-political levels for seed system rehabilitation interventions.

STAKEHOLDER RESOURCE	LEVEL OF ACTIVITY				
	LOCAL	DISTRICT	NATIONAL	REGIONAL	INTERNATIONAL
1. FARMERS / LOCAL COMMUNITY					
Farmers	*				
Farmer and rural organisations	*	*	*		
Community organisations	*				
Extension agents	*	*			
2. SEED INDUSTRY					
Seed banks	*	*	*	*	
Seed growers	*				
Seed exchange / barter	*	*			
Seed companies / marketers		*	*	*	
Parastatal organisations		*	*		
Seed associations			*	*	*
3. RESEARCH, DEVELOPMENT, EDUCATION					
R & D Institutes		*	*		
Education Institutes		*	*		
IARCs			*	*	
Technical co-operation			*	*	
Regional networks				*	
Gene banks			*	*	*
4. GOVERNMENT					
Local / district administration	*	*			
National authorities / agencies			*	*	*
Regional associations				*	
International conventions / agreements					*
5. DEVELOPMENT / AID COMMUNITY					
Local NGOs	*	*			
INGOs	*	*	*		
Donors – government / philanthropic			*	*	
UN agencies			*	*	*
6. FINANCIAL RESOURCES					
Local micro-credit	*				
Rural finance institutions	*	*			
National finance institutions			*		
Development banks			*	*	*

Table 3. **Proposed strategy to adopt for the restoration of farmer seed systems, plant genetic resources and seed security.**

PHASES IN THE DISASTER CYCLE	ACTIONS TO BE CARRIED OUT
Pre - disaster	* Construct crop genetic resources databases on: - Variety & land race diversity - Agro-ecological adaptation - Biotic and abiotic constraints - Socio-economic characteristics * Effective use of early warning information
During disaster	* Needs assessment * Procuring * Multiplying * Distributing
Post - disaster	* Impact Assessment / need for further actions

Table 4. **Recommendations for the implementation of the proposed strategy for restoration of farmer seed systems, plant genetic resources and seed security.**

PHASE	RECOMMENDATION
Pre-disaster	1. Establish a mechanism that will: a) obtain input from seed security stakeholders; and b) facilitate making information on CGR and seed systems readily available to stakeholders.
During disaster	2. Prior to seed procurement, in an emergency response, FAO explores possibilities to support seed security needs assessments of stakeholders involved in the emergency response. 3. FAO establishes a mechanism to encourage and co-ordinate regional harmonisation of seed rules and regulations. 4. FAO promotes a mechanism to co-ordinate procurement, multiplication and distribution of quality seed and planting material of adapted varieties at the national and/or regional level and to integrate with local seed distribution systems.
Post-disaster	5. FAO encourages assessment of impact of interventions to rehabilitate agricultural systems after disasters (disseminate lessons learned – what worked well and why; what did not go well and why not).

DEVELOPING INSTITUTIONAL AGREEMENTS AND CAPACITY TO ASSIST FARMERS IN DISASTER SITUATIONS TO RESTORE AGRICULTURAL SYSTEMS AND SEED SECURITY ACTIVITIES

Introductory Remarks

by

Dr. A. Sawadogo
Assistant Director General
Agriculture Department
FAO, Rome

Distinguished Colleagues and Friends.
Greetings and a warm welcome to all of you from FAO.

The development and sustainability of agriculture depend on the conservation of plant genetic diversity. The access of farmers to diverse local varieties has been, and will continue to be, the foundation of present and future agricultural systems in developing countries. Therefore, the protection of local varieties and the availability of seed of adapted varieties is critical to efforts to achieve seed security and food security.

Many farmers in developing countries still largely depend on on-farm production of local seed and plant material. These agricultural systems have been time-tested and are found to be stable. However, they are also extremely fragile during times of disasters.

The incidence of disasters caused by events such as war, civil strife, drought, flood, and fire has increased in the past decades and is at record levels in the 1990s. Such calamities have profound effects on agricultural systems, involving farmer seed systems as well as the social structures which bind communities together. In the wake of a crisis, farmers' seed stocks can be destroyed, stolen or consumed, and seed of locally adapted crop varieties may be lost entirely. During disasters, countries and communities are often caught off-guard, and they frequently lack the capacity to restore their seed supply systems, often leading to long-lasting seed and food insecurity.

Common national and international responses to calamities have been in the form of emergency relief activities, including seed supply for planting to restore agricultural systems in the affected areas. Emergency seed relief programmes have helped millions of farmers in different developing countries. However, due to the complexity and difficulties related to emergency operations, these programmes have been, in most cases, unable to restore the majority of the adapted crop diversity when lost in a region. The potential negative effects of the introduction of inappropriate varieties, including yield reduction, may be long-lasting. Moreover, the diffusion of untested material may pollute germplasm of the remaining local varieties, thus accelerating genetic erosion.

These difficulties were recognised by the representatives of 150 countries at the 4th FAO International Technical Conference for the Conservation and Sustainable Utilization of Plant Genetic Resources for Food and Agriculture, held in June 1996 in Leipzig, Germany. During the Leipzig conference, the Global Plan of Action (GPA) was adopted. Activity number 3 of the GPA calls for assisting farmers in disaster situations to restore agricultural systems. The coordination of this activity was mandated to the FAO in close collaboration with WFP, UNHCR, UNDRO, IPGRI, national and international agricultural research institutions, countries affected, donors and Non-Governmental Organizations.

Seed security for food security is a major priority of the FAO, to which the Director General attaches particular importance. In this regard, the FAO and the Accademia dei Georgofili, the Italian Academy of Agricultural Science, jointly organized the International Technical Meeting on Seed Security for Food Security, in 1997, in Florence, Italy. During this conference, the importance of developing mechanisms to facilitate the restoration of local crop varieties after disasters was stressed as being critical to the re-establishment of seed supply and agricultural systems.

Conscious of the high costs of emergency operations, FAO continuously seeks to prevent disaster-related emergencies. When they do occur, FAO's goal is to reduce their impact and to accelerate a recovery process that will lead to sustainable agricultural development.

FAO has invited all of you who have been working towards seed relief and rehabilitation of plant genetic resources after disasters to share lessons learned. I am confident that, through this workshop, strategies will be identified to improve the capacity of disaster-prone countries and their rural farming communities to identify, acquire, multiply, and deliver seed of locally adapted varieties before, during, and after disasters. The workshop also has the aim to determine mechanisms of collaboration for a continuing partnership among governments and institutions involved in the maintenance or restoration of local plant genetic resources and seed supply systems.

I would like to thank the Seed and Plant Genetic Resources Service (AGPS) of the Plant Production and Protection Division of the FAO for organizing this workshop. My gratitude goes also to the International Plant Genetic Resources Institute, the Special Relief Operations Service of the FAO and the World Food Programme for their collaboration in the preparation for this important meeting.

Please let me take this opportunity also to thank the Government of Norway for supporting this project.

Before I end, please allow me to extend the greetings of the Director General of FAO, Dr. Jacques Diouf, to all of you. I wish you a pleasant and fruitful meeting.

Thank you for your attention.

SEED AND PLANT GENETIC RESOURCES RESTORATION IN DISASTER AND CONFLICT SITUATIONS IN ANGOLA: SOME EXPERIENCES FROM OVER 20 YEARS OF CONFLICT SITUATIONS

Elizabeth M. Matos
Chairperson, National Plant Genetic Resources Committee, Luanda, Angola

1. INTRODUCTION

In October 1998, Angola is still in a serious phase of increasing internal conflict, with all this implies in terms of death, destruction, insecurity, restricted movement of people and produce. While the country is still unable to implement measures for a post-conflict situation, it has been in a high or low level conflict situation for over 20 years and cannot afford to wait for its final conclusion before taking measures to ensure local seed supplies and to conserve its great wealth of plant genetic resources (PGR). Over 85% of Angola's farmers are small-scale subsistence farmers.

Angola has an area of over 1,200,000 km^2, ranging in latitude between approximately 8° S and 17° S and in altitude from sea level to over 1,500 m. Its natural vegetation includes humid tropical forests, savannas and deserts, and it has climates and soils that support a large range of tropical and temperate crops. Its population is estimated at between 12 and 13 million.

The nature of the conflict in Angola in relation to seed supplies and PGR conservation.

Angola has been in a state of internal armed conflict since before independence in 1975. The great majority of the population lives in government held areas. This includes most of the urban population and large agricultural areas, although at various times over the past two decades in some provinces these have been restricted to a few tens of kilometers around provincial capitals, and with high populations of displaced people. The highest population centres are in coastal towns (3 million in and around the capital) and escarpment areas to the east of the coast. Apart from regional drought episodes, these areas are prone to insufficient and/or erratic rainfall.

In the period between late 1970s and 1992, seeds were imported through the sole government seed importing company, "Angosementes". The Ministry of Agriculture distributed these seeds in government areas. Until 1980s, there was small-scale multiplication of basic seed produced at the Agricultural Research Institute in Huambo.

Unita held areas have been mainly rural, inland, at higher altitudes and with more reliable rainfall patterns. These are generally better agricultural areas and over the period up to 1992 received very restricted supplies of imported seed. As a result of restrictions on movement and consequent isolation of many farming communities in both government and Unita held areas, a myriad of local varieties have emerged, from what was effectively an enforced *in situ* conservation situation.

The most destructive conflicts occurred in the period immediately following nationwide elections in 1992, causing further widespread population displacement (up to one quarter of the whole population) moving mainly to the relative safety of government held provincial capitals and principally to the drier to semi-arid coastal provinces. A new cycle of conflict and consequent displacements has arisen since May 1998.

2. PLANT GENETIC RESOURCES

Prior to independence there was no recognition in general, nor in the plant breeding department of the Agricultural Research Institute, of the need to conserve plant genetic resources *per se*. Breeders' held working collections of commercial crops. There were no plant breeding programmes for main subsistence crops of local farmers (e.g. sorghum, millet, and cowpea) and thus no *ex situ* conservation of PGR of local farmers' varieties of subsistence crops. As a result of the size and variation in country's physical/climatic conditions there was, and still is, a vast wealth of local farmers' varieties, almost all in farmers' fields. Extremely few Angolan PGR were collected and placed in international banks in the colonial period.

In the post independence period, due to the continuous war situation, no collections were made by international institutions. There was also no national recognition of the importance of PGR conservation until 1987 and the setting up of the SADC PGR Network and the National Plant Genetic Resources Committee in 1989. There are now some 1,200 accessions of local crop varieties in the National Plant Genetic Resources bank, collected since 1991.

On the down side, the 800 accessions held at the Agricultural Research Institute (ARI) in Huambo, including a large number of breeders lines and some local varieties, were lost during the period of Unita's occupation of Huambo between 1992 and 1994.

Angola's rich agricultural germplasm is not, with very few exceptions, being conserved in international gene banks. Thus what is, or has been, lost as a result of disaster and conflict, cannot be re-introduced or regenerated, from internationally held material.

3. SEED AND TOOL DISTRIBUTIONS TO DISPLACED PEOPLE AND RETURNEES (POST 1992)

Following the cease-fire in 1994, Angola has experienced over 3 years of efforts to implement the Lusaka Peace Accord, and at the time of writing there is a serious upsurge in conflict. Between 1995 and 1998 international UN agencies, the government, international and national NGOs made emergency distributions of seeds and tools to displaced and war-torn populations, and post 1996, to those beginning to return to their home areas.

The results of four evaluation surveys of these distributions, made between 1996 and 1998, covering a total of more than 3,000 farmers in 13 provinces, together with the experience of 7 years of PGR collecting in the accessible parts of the country, are the basis for the following conclusions.

Although farmers are keen to receive new and improved seed, they are well aware of the adaptive characteristics of their own local varieties, especially in marginal areas and with low inputs. Farmers make every effort to hold onto their own seed stock, the variety whose characteristics they know and trust. A very high proportion (over 90%) of displaced and resident farmers sowed at least some local seed varieties. Some of this seed in recent years has undoubtedly come from Unita held areas, the result of exchanges made for imported goods from coastal areas. Traders, often women taking lifts on variety of precarious means of transport, will cross enemy lines to barter imported goods for grain of various kinds.

The surveys showed that displaced farmers did have local seed, obtained from resident local farmers, from markets, and other sources but they may also have lost the local variety they know or would like to plant. It has not yet been possible to identify which varieties have really become extinct and which may still be available in areas that are inaccessible at present.

While emergency seed supplies are required in specific drastic situations there are disadvantages to the blanket distribution of imported seed. It is often not appropriate to local conditions, specifically to the marginal, low-input management conditions of displaced farmers.

In highly variable areas with, for example, differing growing season lengths, no single variety will be appropriate for all areas.

In conflict and disaster situations, imported seed is often late arriving in the hands of the displaced farmer. This is crucial if the crop demands that seed is sown at first sign of short rainy season (more than 75% of farmers surveyed requested earlier seed deliveries).

4. REQUIREMENTS FOR INCREASING CROP AND SEED PRODUCTION.

Displaced and returning farmers need to be helped to cultivate a larger area. They may have lost their animal traction/ploughs and need new animals and ploughs (or they need some one with a tractor to plow and prepare the area to be sown).

They need hand tools (such as hoes, cutlass and file), and in sufficient quantity that several members of a family can work in the field simultaneously.

In Angola, the need for seed was not the first priority of displaced or returning farmers, since in most areas they could obtain seed one way or another. The exceptions to this are when a large displaced population seeks refuge entirely within a 5-10 km radius of the provincial capital, e.g. Malange 1993/4. In such cases, there can be a total lack of seed. In these situations, a much higher amount of seed, preferably from the nearest possible sources, should be distributed to displaced farmers.

Farmers showed a strong desire to receive improved imported material, to try it out in their own conditions, to allow crossing and selection to improve their locally adapted varieties. It would be highly recommend for the wide scale introduction of participatory plant breeding projects, particularly for farmers in difficult marginal situations. This would also mean the opening up of genebanks for such use. However, to do this, there is a need to establish an area in the genebank (or elsewhere?) that is responsible for the multiplication of genebank material for distribution. In Angola at present, only samples of 50-100 seeds maximum are available for distribution. The NPGR programme is just starting a programme to multiply some accessions for wider distribution to farmers and NGOs.

Before introgression has altered some local characteristics and before other varieties are lost through abandon in times of conflict, emergency collection of PGR is needed in conflict and disaster areas. The means used to do this may not be conventional PGR collection missions. Local staff, NGOs, extension workers can be trained as "paracollectors" to do this urgent task. In Angola, national and provincial workshops have been held to raise awareness of the value of local PGR, maintenance of agro-biodiversity and to train participants in methods of small-scale seed production, improved local seed storage and the collection of local germplasm. The result is that the national bank has now received numerous samples from areas that are still inaccessible to its staff. These last activities have been possible through a FAO/TCP project to save local germplasm.

There is need for safety duplicate collections to be held in a bank in another part of the country, outside the country, and/or in regional or international bank for safe-keeping.

The National PGR Committee is unequivocal that until such time as there is international agreement on the fair and equitable sharing of benefits from PGR, international NGOs should be under an obligation to abide by the International Code of Conduct on the Collection and Transfer of Plant Genetic Resources. They should not remove PGR from the country in a disaster situation, at least not without the prior informed consent of the country concerned.

In the area of seed multiplication, there is a recent example of successful collaboration between local farmers and the ARI in Huambo that has resulted in the multiplication of a good local variety of maize. This was purified by breeders in the ARI and small quantities distributed to farmers associations. They multiplied the seed, returning the original amount to the project for distribution to other farmers and divided the rest of the yield between members of the association. By the end of the season, other farmers were queuing to join the scheme.

Case Studies

AGRICULTURAL RECOVERY AND EMERGENCY SEED RESTORATION IN THE POST DISASTER SITUATION OF ANGOLA. A CASE STUDY: WORLD VISION INTERNATIONAL

Claude Nankam
Director, Agriculture Programme
World Vision International
Luanda, Angola

Abstract

After three decades of civil unrest, the population of Angola was completely devastated. People lost their belongings as they were fleeing from the war. With the consolidation of the peace, Private Voluntary Organizations (PVOs) were faced with the tasks of improving health and nutritional status of resettling communities. Seeds and tools were distributed alongside food and health assistance, as integral parts of the resettlement of returning, internally displaced and impoverished people. The destabilizing environment was characterized by discontinuity in agricultural technology research and development systems and the loss of information on appropriate cropping systems and varietal suitability for small-holder farming. This paper discusses initiatives undertaken by World Vision International to rehabilitate agricultural production of the Angolan small-holder farm family, and to restore a sustainable seed system within communities.

1. INTRODUCTION

War, drought, economic and social displacement, and disruption weaken and destroy mechanisms for achieving food security and maintaining family livelihoods. After three decades of civil unrest, along with natural disaster such as drought, the population of Angola in 1993, was made up of 63.7% war-affected people, 17.5% displaced people, 13.1% drought affected people, and 5.7% people returning to their land of origin (UN Report, 1993).

In 1993, World Vision International (WVI), with substantial private and government donor support, launched a multi-sectoral developmental relief and transition programme designed to address short and medium term needs of war-affected populations in Northern Angola, principally in Kwanza Norte, northern Kwanza Sul, and Malange Provinces. Short term interventions included the challenging tasks of improving health and nutritional status of resettling communities. Food aid was distributed to mitigate starvation. Feeding centers were created for the rehabilitation of mal-and/or under-nourished children and mothers. Mobile clinics and health posts were established to improve health of the population. Agricultural programmes were developed to address food shortages. This paper presents some of the key aspects of this experience pertaining to the recovery of agricultural production both short and medium terms, leading to food self-sufficiency of affected population, as a means of addressing issues of poverty and hunger and longer term with a view toward restoring economic stability and laying the ground work for future economic and agricultural development. Aggravating household food insecurity is the unavailability or short supply of seeds and planting materials, the discontinuity in agricultural technology research and development systems and the loss of information on appropriate cropping systems and varietal suitability for small-holder farming (Chapman *et al.*, 1997). World Vision's experience in supplying quality seeds within this destabilizing environment is described, followed by lessons learned.

2. AGRICULTURAL RECOVERY

During the 1993-1998 period, with major support from USAID, DFID, AusAID, CIDA, UNICEF, UCAH, WFP, WVI private funds and other donors, World Vision played an important role in facilitating resettlement and rehabilitation by assisting returnees and vulnerable populations in their efforts to reestablish mechanisms for attaining a minimum level of food security and access to basic health care services through the provision of direct technical and commodity assistance. For Chapman *et al.* (1997), the following situations characterized the problems encountered by World Vision at the initiation of the agricultural recovery process:

- Stocks of seed and planting material were disrupted, lost or in short supply. This required an emergency injection of planting materials from outside sources.

- Returning refugees and displaced impoverish populations lacked the means to purchase agricultural inputs. Seed "paks" were therefore distributed free of charge, along with food rations to tide people over until they could harvest crops and begin to feed themselves.

The success of helping the small farm family achieve food security depends significantly on which varieties are included in the seed "pak" (Tripp, 1998). Unfortunately, no varietal screening was done. Therefore, the seed "paks" often included whatever seed was generally available (Chapman *et al.*, 1997). In some cases, grain was bought and shipped as seed (Rohrbach *et al.*, 1997), or a variety was shipped with the commercial name of another variety, e.g. the groundnut variety "Spanish" was shipped as Natal Common, although Natal Common is no longer produced as seed (Hildebrand, personal communication).

A "pak" consisted of one variety of each of the main staple food crops: maize, beans, groundnut, and cowpea. These "one size fit all" seed "paks" were distributed by PVOs throughout the country, without regard to environmental factors such as elevation, rainfall, temperatures, etc. This distribution of largely inappropriate seeds varieties had the effect of retarding the pace of agricultural recovery because many varieties in the "paks" performed poorly in Angolan environments for which they were not adapted.

In 1996, World Vision designed a "farmer first" extension service and modified the composition of seed "paks" to include two adapted maize varieties so that recipients of seeds and tools will obtain optimum production from the farm inputs they received. Table 1 presents the number of farm families per year who received seeds and tools and benefited from World Vision extension services.

During the 1996/97 growing season, a Food For Agriculture (FFA) initiative was launched whereby food was used as an incentive to increase cultivated land. A farmer was given a food ration when he had cultivated a minimum of 2500 m^2 of land. Through this initiative, cultivated land in 1997/98 increased by 43 % in Malange Province and more than doubled in Kwanza Norte, compared to 1996/97 production.

World Vision programmes emphasized formal training of technicians, extension agents and leading farmers through seminars and workshops. On-the-job training of farmers was a key component of the extension activities. Training was done during visits to farmers' fields, and during farmers' field days organized at demonstration plots. Primary recommendations on farming systems included:

- **Plant density:** A survey on the rate of adoption of recommendations for appropriate plant density showed that in Kwanza Norte 54.7% of farmers were using appropriate plant density, while in Malange only 19.4% had adopted the recommended plant density. Quantitative data from demonstration plots in Malange showed 18% increase in yield when WVI recommended densities were used, compared to plant densities selected by farmers.

- **Erosion control:** On erosion control, only 22.8% of farmers in Kwanza Norte cultivated their land using contour ridges (perpendicular to the slope), to prevent erosion of top soil by rains. In Malange, 54% were controlling appropriately soil erosion as recommended by the extension team.

Table 1. Agricultural and vegetable seeds and tools distributed by World Vision from FY94 to FY98

Financial Year	Number of Families	Total beneficiaries*
1993-1994	7,963	29,815
1994-1995	25,482	132,410
1995-1996	63,518	317,590
1996-1997	77,000	385,000
1997-1998	50,852	254,260

* Total beneficiaries is estimated with the assumption that the average family size is five people.

Table 2. 1997/98 production estimates for beans, cowpeas, maize, and groundnuts in WVI project area

Province	Munici-pality	Beans (kg/ha)	Cowpeas (kg/ha)	Maize (kg/ha)	Groundnut (kg/ha)	Estimate 1996/97(kg/ha)	
						Beans	Groundnut
Kwanza Norte	Cazengo	1496	1273	916	430	801	542
	Lucala	436	387	1246	1075	419	703
	G. Alto	1755	1247	1700	928	n/a	n/a
	S.Caju	329	800	n/a	n/a	405	341
Kwanza Sul	Libolo	1393	n/a	1165	n/a	n/a	n/a
Malange	Malange	545	n/a	6532	n/a	215	n/a
	Cacuso	743	n/a	n/a	n/a	n/a	n/a
	Mucari	863	n/a	5741	n/a	n/a	n/a
	Quela	n/a	n/a	5717	n/a	n/a	n/a

n/a: Not available.

Due in large measure to WVI's current programme strategy, most assisted villages have achieved at least a basic level of food production sufficient to provide for family food needs for 6 to 10 months. Table 2 shows World Vision yield estimates for 1997/98 growing season for beans and groundnut compared to those in 1996/97. For maize, FAO estimates of production in 1996/97 were 410 kg/ha, 530 kg/ha, and 510 kg/ha for Kwanza Norte, K. Sul, and Malange Provinces, respectively, and are used to assess the 1997/98 production.

These figures indicated that maize production in 1997/98 was up 220% to 415% in Kwanza Norte, and 220% in Kwanza Sul over 1996/97 production levels. Bean yield estimates for 1997/98 compared to 1996/97 production ranged from 81% (a decrease probably due to excessive rain in the micro-climate of Samba Caju) to about 186% (Cazengo) in Kwanza Norte, to about 300% in Kwanza Sul (using FAO 1996/97 estimates), and from about 182% to about 250% in Malange (using WVI and FAO 1996/97 estimates).

To sustain this increased production, WVI designed the "Seeds of Freedom" Project (SOF) to provide quality seeds and planting materials to farmers. The next section presents the SOF approach for seed restoration and preliminary results.

3. SEED RESTORATION IN EMERGENCY SITUATIONS

3.1. On-station and on-farm testings

Through an array of seven Private Voluntary Organizations (AFRICARE, CARE, CONCERN, CRS, LWF, SCF(US), and WVI), the Institutes of Rural Development (IDA) and of Agronomic Research (IIA) of the Ministry of Agriculture (MINADER), farmers were given the opportunity to test, on their farms and under their own growing conditions and management, improved technologies developed by five International Agricultural Research Centers (IARCs): CIAT (for beans), CIP (for Irish potato and sweet potato), CIMMYT (for maize), ICRISAT (for sorghum, pearl millet and groundnut), and IITA (for cassava, and cowpea). In these on-farm trials, farmers compared new varieties to local varieties. Farmers selected varieties base on criteria such as yield, cooking and palatability tests, milling properties, etc. The selected varieties were reported to the National Seed Service (SENSE) by IIA and IDA, to be incorporated into the list of varieties recommended for specific agro-ecologies of the country. Angola was divided into six distinct agro-ecological zones, using six weather and soil parameters, including elevation, rainfall, maximum and minimum temperatures, evapo-transpiration, and soil water holding capacity (Fig. 1).

During the 1996/97 and 1997/98 cropping seasons, respectively 1,040 and 1,950 on-farm trials of beans, maize, sorghum and pearl millet were established in five of the six agro-ecological zones. Three improved varieties of beans (CAL 113, A197 and A286), one variety of pearl millet (Okashina 1), a local variety of sorghum, two improved varieties of maize for the mid-altitude zones (ZM 607 and Matuba) and three local varieties of maize (SAM 3, Branco Redondo and Dente de Cavalo) for the high altitude zone (Central Plan-Alto) were most preferred by farmers over the other varieties. Three clones of cassava (two improved lines TMS 30211 and TMS 42025 and one local "Precoce de Angola") previously selected by farmers were in the multiplication scheme for wide distribution to farmers.

Figure. 1: Main Agroecological Zones in Angola

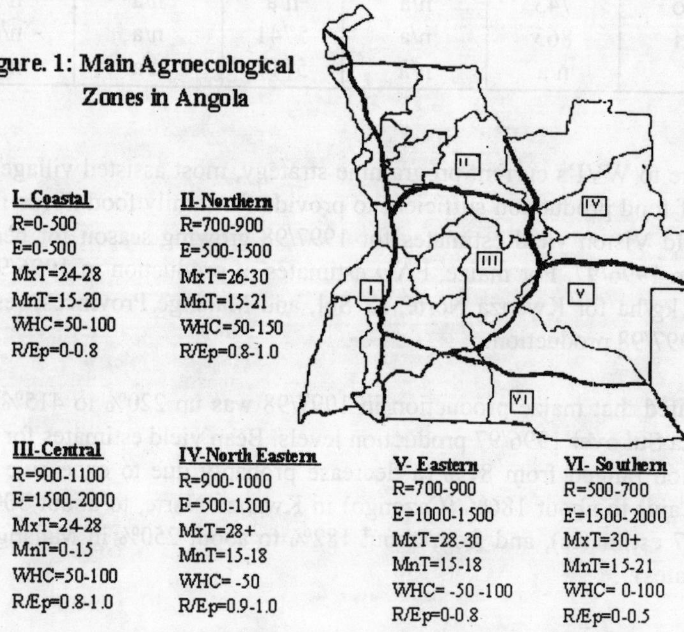

Tables 3 and 4 presents the yield advantage of farmers' selected varieties over the local varieties. Table 5 summarizes varieties of beans recommended by the project for each agro-ecological zone.

Table 3: Yield advantage (%) of the most preferred bean variety in comparison to local varieties

Agroecological Zone	Yield Kg/ha		Yield advantage (%)
	Preferred	Local	
Zone I	846	617	62
Zone II	333	267	25
Zone III	327	156	112
Zone V	657	443	15

Table 4 : Yield advantage (%) of the most preferred bean variety based on palatability, in comparison to local varieties

Agroecological Zone	Yields Kg/ha		Yield advantage (%)
	Preferred	Local	
Zone I	325	217	40
Zone II	333	267	25
Zone III	327	156	22
Zone V	535	443	17

Table 5. Varieties preferred by farmers in each agro-ecological zone

Agro-ecological Zone	Varieties (ranking)
Zone I	A 344 (1st), Local (2nd)
Zone II	A 197 (1st), A 344 (2nd)
Zone III	A 286 (1st), CAL 113 (2nd)
Zone V	A 286 (1st), A 344 (2nd)

3.2. Seed multiplication and production

Multiplication of foundation seeds by IARCs

For the 1996/97 season, seeds were shipped from the Centres in kits for on-farm trials, and in bulk for distribution to farmers. Planting materials for cassava, potato and sweet potato, as well as True Potato Seed (TPS) were sent by IITA and CIP for increase and trials on-station:

- 15 tonnes of pearl millet (2 varieties) and 80 tonnes of sorghum seed (3 varieties) by ICRISAT;
- 4 tonnes of beans (4 varieties) by CIAT;
- 14 tonnes of maize (13.7 tonnes of one intermediate maturing and 300 kgs of two late varieties) by CIMMYT;
- 15,000 pathogen-free plantlets of 16 varieties of cassava regenerated from tissue culture from IITA;
- 380 pathogen-tested cuttings of 17 high performing sweet potatoes varieties for testing by IIA and WVI, 1460 minitubers of 9 advanced potato clones; and
- a packet each of 1000 true potato seed (TPS) of 6 hybrid progenies, shipped from CIP, to be tested by IIA and CONCERN in Huambo.

Multiplication of farmers' selected varieties

Large scale seed production: Eleven tonnes of seed of the maize variety ZM 607 were produced by an orphanage, "Casa de Gaiato" in Malange, in a partnership with World Vision, where the orphanage provided World Vision a testing site, and World Vision trained the orphanage in farming systems and seed production techniques. As a result of this joint venture, the orphanage now is able to produce sufficient grain for its 170 children, and to generate funds by producing and selling good quality seeds. Medium to large scale farmers also are being trained in order to develop a local seed production system.

Production of root and tuber planting materials: As a step toward commercial seed and planting material production, World Vision contracted Sociedade Agro Alimentar Lda (SAA), a private company operating a 1400 hectare farm in Catete, Bengo Province, to produce and deliver to World Vision 450,000 "cassava cuttings" for distribution to 50,000 farmers and SOF NGOs in 1999. To do this, the farm received technical advice and on-the-job training during field visits from World Vision agriculture staff. The farm is fully equipped with an overhead pivot irrigation system and adequate farm machinery and equipment. Another commercial farm, Kalakala, was contracted to produce 60 tonnes of the maize variety "Matuba" and 30 tonnes of bean varieties A286 and A344 for distribution to farmers (Table 6).

Seed production by small farm holders: In September 1997, seeds of maize and bean varieties selected by farmers during the 1996/97 on-farm trials (ZM607 and Matuba for maize, Carioca and A286 for beans) were distributed, on a loan basis, to farmers in the World Vision project areas. After harvest, each farmer reimbursed to World Vision the same amount of seeds he or she had received.

As indicated in Table 7, about 30 tonnes of maize seeds and 27 tonnes of bean seeds were received from farmers. These were distributed as uncertified seed to farmers in other communities which represents 33% of the total seed of maize and beans distributed by World Vision in September 1998. This informal seed production system will continue to be used for seed multiplication. Farmers will receive training in seed production technology in order to produce good quality seeds that will be certified by SENSE and distributed as foundation seeds in other agro-ecological zones where they have been chosen by farmers. During 1998/99 season, all SOF stakeholders will be encouraged to coordinate their seed activities through the SOF coordinator and collect data on the adoption rate as well as seed production of the farmers selected varieties.

Table 6. Production of commercial seed and planting materials in Angola

Crop type Variety	Area (ha)	Expected yield (ton) or cuttings	Remarks
Beans A286	20	20	
Beans A344	10	10	
Maize (Matuba)	40	60	
Cassava	42	2,500,000 cuttings	50 cuttings per farmer X 50,000

Table 7. Total quantity of seeds reimbursed to World Vision by farmers as pay back for seed loan for the 1997/98 growing season

Municipality	# families that received seed loan	Quantities of seeds reimbursed (kg)		
		Beans	Cowpeas	Maize
Kwanza Norte	18644	609 (1.1%)	1566 (2.8%)	4940 (8.8%)
Kwanza Sul	6623	8804 (44.3%)	7746 (39%)	12000 (60.4%)
Malange	17932	18227 (33.9%)	577 (1.1%)	13000 (24.2%)
TOTAL	43199	27640 (21.3%)	9889 (7.6%)	29940 (23%)

World Vision conducted a food security baseline survey in August 1997 which indicated that 21% of farmers saved their own seeds, while 48% bought seeds from the local markets, 33% received seed from WVI, and about 7% could not plant due to lack of seeds. Only 2,990 farmers were involved in the on-farm trials, although a total number of 10,663 farmers participated in different tests during the field days. The implication of this is that few farmers had access to seeds of selected varieties. The need for seeds of cereals (maize, sorghum and millet) and beans for Angola in the 1998/99 growing season are estimated at 10,000 tonnes (FAO/WFP Food Assessment Mission, May 1998) and in WVI areas, the estimate is 1,300 tonnes. Will this quantity of seeds be available? If yes, what percentage of these will be farmers' selected varieties?

It is clear that seed saved by farmers cannot meet this level of demand. Thus, there is need for developing a strategy for seed production and distribution that goes beyond seed saved by farmers and movement of seed between farmers at the local level.

3.3 Seed delivery systems

Tripp (1998) described seven options for seed delivery. Two of the options will be examined by SOF as possibilities for developing an Angolan national seed system.

Co-production by farmers in "pilot villages"

The term co-production is used here to imply that varieties are developed by public research services, but the farmers play a role in multiplying and distributing seed of these varieties (Tripp, 1998). World Vision has identified farmers and farmers groups in some communities ("pilot villages") who can multiply seed of farmers' selected varieties of bean and maize. The success of this initiative will depend on farmers training in seed production technologies, and incentives in terms of seed storage and marketing. The key issue is the introduction in the community of seeds of preferred varieties, to initiate farmer-to-farmer seed movement, while promoting excess production that NGOs will buy for emergency seed distribution programmes. The role of the research institution and the national center for plant genetic resources is crucial to maintain reliable sources of varietal purity for seed replacement.

Small scale seed enterprises

Government regulations on seed importation are increasingly limiting seed importation. Many of the farmers' selected varieties are not yet being produced by large scale seed companies outside of Angola. World Vision has identified medium to large scale farmers who are capable of producing a large portion of the Angolan seed needs, if these farmers could embark into the seed production business. The key issue, and this is also true for the pilot village seed production, is the introduction of low cost seed production techniques to keep seed prices as close as possible to grain prices, thus enabling farmers to buy the locally-produced seed. As the variety delivery system is developing, certification will become important with suitable incentives that will improve prices.

4. LESSONS LEARNED

4.1 Positive aspects of the experience

- Emergency situations break down institutional barriers to agricultural technology transfer;
- Effective strategies for disaster mitigation re-establishes family productivity and livelihood;
- Participatory approaches, with farmers heavily involved in the process from the beginning, helps ensure high adoption rates and programme efficiency;
- Lower barriers to seed-embedded technology adoption providing suitable evaluation methodology and evaluation criteria are used; and
- Highly cost effective and inherently sustainable as it shortens the time that food aid is needed.

Value of partnerships among different organizations are essential, each playing the role for which it is best suited, such as IARCs providing candidate varieties, seed companies providing seed and packaging services, farmers screening varieties and providing feedback on best varieties, and NGOs which facilitate the process.

4.2 Issues and constraints

- Initial unavailability of seeds of adapted varieties implies that the programme is not as effective as it could be if improved varieties were tested and readily available;
- Distribution of grain as seed creates a lost opportunity and quality control problems, a cheap and quick option that in reality is quite expensive; and
- Farmer habit of saving seed inhibits market development

5. LITERATURE CITED

Chapman, J., White, J., and Nankam, C. 1997. World Vision's experience with seed suppiy during emergency and resettlement programmes in Mozambique and Amgola: Implications for the future. Paper presented at the ICRISAT/ICARDA/IITA/GTZ Workshop, "Enhancing Research Impact Through Improved Seed Supply: Options for Strengthening National and Regional Seed Supply Systems", Harare, Zimbabwe, 10-14 March 1997

Tripp, R. 1998. Seed systems and the delivery of new crop varieties. Paper presented at the Sorhum and Pearl Millet Improvement Programme Stakeholders Conference. Harare, Zimbabwe, 27-30 July 1998.

Rohrbach, D., Mutiro, K., and Mazhzngara, E. 1997. Seed availability and markets: the case of sorghum and pearl millet seed supply in Zimbabwe. Pages 21-31 *in* Proceedings of the Zimbabwe National Sorghum/Millet Programme Workshop, 18-19 Feb 1997, Harare, Zimbabwe (Mangombe, N., Heinrich, G. M., and Gono, L. T., eds.). Bulawayo, Zimbabwe: Sorghum/Millets Team, Department of Research and Specialist Services.

Case Studies

EMERGENCY SEED SUPPLY IN AFGHANISTAN

N. S. Tunwar
FAO Senior Technical Adviser (Seed)
Herat, Afghanistan

Executive Summary

This paper describes a case study of the Seed Programme being implemented since 1988 in Afghanistan weakened already by years of war, and further stripped of personnel after the collapse of the communist government.

Afghanistan is a small, land-locked country and its strategic importance lies in its extensive border with various countries and its proximity to the Gulf. The climate over most of the country is arid and semiarid with rainfall ranging from 100mm to 400mm. Orogenic processes are still active as evidenced by time to time seismic vibrations taking place in Afghanistan. Recently, two major earthquakes hit Badakhshan and Takhar provinces killing more than 4,000 people and causing large scale damage to the property. The total area of the country is 65.2 million hectares and only 7.6 million hectares or 11.6% is cultivable. The total area under annual crops is approximately 3.7 million ha i.e. 5.6 % of the total land area.

The war has had a devastating effect on Afghanistan. It has killed at least a million people, maimed and disabled many more and created an army of orphans and widows. Agriculture has been the hardest hit by the perpetual internal conflict. By 1988-89, the total crop area declined by more than one-fourth (28%) and the aggregate production in real terms has fallen nearly by as much (25%). One of the war's legacies is that, although food production was seriously reduced, growing of poppy flourished to such an extent that Afghanistan ranked first. Once thriving villages have become desolate ruins of mud; orchards, vineyards and terraced hillside fields are dead or overgrown; and farms are littered with millions of mines.

The importance of good seed was recognized in Afghanistan in early 1970`s. The Afghan Seed Company (ASC) was established in 1976 as a Government Joint Stock Company to help in developing seed industry in the country. Later its name was changed as Improved Seed Enterprise (ISE). The Headquarters of the Enterprise was in Kabul and it had several sub-offices in different provinces. The UNDP/FAO helped ISE through a Seed Production and Training Project, AFG/86/010 and established various facilities. The ISE functioned as an effective counter part in government controlled area up to April 1992. At the time of collapse of Dr. Najibulah government most facilities were lost due to large scale looting which followed the change in the regime. Following the signing of the Geneva Accord in 1988, a second focal point based in Pakistan was opened for across-border areas not controlled by government.

The programme was funded through the United Nations Office for the Coordination of Humanitarian and Economic Assistance Programmes for Afghanistan (UNOCA), now called UNOCHA. A number of international donors and UNDP supported the programme. Major contribution was received from the Governments of the Netherlands, Japan, Canada, Switzerland, Norway and the EEC.

The main aim of the seed programme from its onset in 1988 was to create a situation in rural Afghanistan conducive to the return of refugees and displaced families and start to redress the damage wrought by the war and assistance for agricultural rehabilitation. Food production and security are the essential and fundamental starting point and seed security and food security comes one after another in the sequence of an agricultural rehabilitation programme. The FAO seed programme sought to expand and sustain the use of improved seed through various activities - on-farm trials, selection, production, outside procurement, in-country seed production, networking, distribution, and popularization of certified seed. Fertilizer was also distributed to maximize the benefits offered by the improved seed. It was not very easy to implement the programme under an ever-changing security and political situation. Such complex problems called for an imaginative implementation methodology of diversification and sustainability to remedy the situation rather than pursuing a typical programme. A vigorous and flexible calendar of operations for all stages had to be drawn up in advance and vigorously implemented. Even economic principle which dictates production of seed in a compact area was deliberately ignored to reach more farmers in isolated and inaccessible areas. The decentralization of the seed production was preferred. A diffuse approach consisting of large number of small seed producing pockets was adopted to ensure that farmers of every region and every stratum of society could benefit from the programme.

The FAO Programme has established in pragmatic ways that the provision of high quality seed and planting material (SPM) to Afghan farmers is one of the most cost-effective and economical forms of emergency aid. An integrated approach has been developed for marshalling, delivering and monitoring inputs. An informal back-up support network has been developed by the Programme. This encompasses the public and private sectors. Collaboration with another UN agencies proved very fruitful. Food-for-seed was designed in collaboration of the WFP. The wheat seed was exchanged at a rate of 1:1.25 for wheat grain supplied by the World Food Programme. This ensured that seed growers had enough food to feed their own families and that a large proportion of their seed crops would travel out of the production areas to other farmers specially in food deficit areas. It also helped to make considerable savings of funds. A Fund Management Committee and a Seed Review Group were constituted to manage funds generated as sale proceeds and to advise on all matters relating to seed. More than US $ two million have been generated as sale proceeds and deposited in a separate account in Pakistan.

A number of constraints had to be overcome. Despite these constraints, more than 97% of the inputs did reach the farmers. By 1997, the Programme had distributed more than 48,413 MT seeds, mainly through NGOs and private voluntary organizations. The programme has been extremely successful and innovative. It continues to play a vital role in the rehabilitation of Afghanistan's agricultural production base after twenty years of conflict. As a result, it has been possible for many thousands of refugees and displaced families to return to their villages in Afghanistan with some expectation of being able to support themselves.

The FAO's Afghanistan seed programme provides many useful lessons. The most significant lesson of all that FAO has learned in Afghanistan is the importance of trusting the individual's own thrust for development. By keeping its intervention small and sharply focused on specific areas and problems it has provided the initial lift that the majority of Afghan farmers in need of before embarking on their own climb upwards to self-sufficiency.

In a world increasingly faced with human tragedies beyond the capabilities of governments to respond effectively, this Programme provides lessons that (with local modifications) will be relevant elsewhere.

1. OVERVIEW OF PROGRAMME / PROJECT ACTIVITIES

General description of Afghanistan

Afghanistan is a small, land-locked country and lies between 29°40' and 38°40' northern latitude and between 60°31' and 75°00' eastern longitude on the mountainous and desert areas where the Iranian Plateau borders with the mountainous systems of Central Asia. Afghanistan's strategic importance lies in its extensive border with various countries and its proximity to the Gulf. These factors have also influenced the pattern of international relations and recent historical events.

The climate over most of the country is arid and semiarid with rainfall ranging from 100mm to 400mm. The country is usually divided into four geographic regions. The northern plains at altitudes of around 400 m have a rainfall of 25 cm per year and provide the main agricultural lands. This area is watered by the Amu Daria and its tributaries. The central mountains reach an altitude of 5,700 m and the snow line in winter creeps down to 2,000 m. The eastern and south-eastern hills and foothills have elevations of 1,300 to 2,300 m and they contain the forest areas of the country. The southern and western lowlands are at altitudes of between 330 and 1,000 m and contain the western desert area of the country where rainfall is as low as 100 mm per year. Through this region flow two major rivers, the Hari Rud and Helmand. Topographical conditions have hampered the development of internal transportation and resulted in the physical isolation of village settlements.

Orogenic processes are still active as evidenced by time to time seismic vibrations taking place in Afghanistan. Recently, two major earthquakes hit Badakhshan and Takhar provinces killing more than 4,000 people and causing large scale damage to the property.

Cultivable area and crop production (Pre-War Era)

The total area of the country is 65.2 million hectares (ha) and due to too steep topography or dry deserts only 7.6 million ha or 11.6 percent is cultivable. The total area under annual crops is approximately 3.7 million hectares, i.e. 5.6 percent of the total land area. Though it is believed that 70 percent of this area is under irrigation but only a small part is under assured irrigation and capable to grow double crops and attain 200 percent crop intensity. Due to unreliable water supply only one crop is grown annually in the major area classified as irrigated. Nevertheless, irrigated area provides roughly 77 percent of all wheat and 85 percent of all food and industrial crops, according to pre-1978 statistics.

The country's main crop is wheat (2.348 million ha in 1978 and 2.124 million ha in 1998). Maize occupies second place (0.482 ha in 1978 and 0.200 ha in 1998) and is grown in the eastern provinces. Barley and rice are also important cereal crops. Rice is mainly grown in the Logar, Laghman and Kunduz valleys while barley and leguminous plants are grown mainly in the mountainous areas. Cotton and oil seeds are produced in the north of the country. Orchards and vineyards can be found all over the country.

Nature of devastation during the war and situation under which the seed programme was implemented

Destruction caused by war was visible every where at the time of launching of the programme. Although no accurate information is available, the destruction caused by the war has been extensive and in some areas of the country extremely severe. Perhaps more than one million Afghans have been killed since 1978; another 250,000-300,000 have been seriously injured (a large number becoming permanently disabled). Over 5 million Afghans became refugees. The country has probably lost 90 percent of its intelligentsia. Some lost their lives but the majority went abroad. How many will return to Afghanistan is an open question. It is estimated that perhaps one half country's population was uprooted at one time or another; the psychological damage to the young and old is said to be considerable. There has been a rapid rise in the rate of illiteracy. It is estimated that 80 percent of the young boys and 90 percent girls are now illiterate teenagers. Lack of food and high prices have reduced the level of nutrition of the population. Tuberculosis has shot up sharply. Most critical is the psychological and emotional disorder of the affected population.

It is the rural areas that have suffered the most severe economic losses. The destruction is not uniform; in some areas the damage has been more severe than others. It is estimated that one-third of the villages in Afghanistan have been destroyed or severely damaged. Heavy damage to the irrigation networks (both modern and traditional like the Karez), destruction of orchards along the main highways as a protection against attacks on military convoys is visible everywhere. Loss of commercial agriculture and resort to subsistence cultivation, leading to considerable decline in yields per hectare is another form of destruction. The country was self sufficient in food production but it has been devastated by over two decades of war. This sector has been the hardest hit by the perpetual internal conflict. By 1988-89, the total crop area declined by more than one-fourth (28%) and the aggregate production in real terms has fallen nearly by as much (25%).

Historical background of the Afghan seed programme

At first, Afghanistan lacked the necessary seed industry infrastructure to make seed available to farmers in sufficient quantity with assured quality. During 1970`s, a significant start was made to setup a state run seed industry, leading up to formation of the Afghan Seed Company (ASC) in 1976 as a Government Joint Stock Company with a view to performing specified tasks as regards the production, processing, testing, certification, distribution and sale of seeds and granting official permits for any sale and distribution of seeds and laying down labeling requirements. Its management was vested in the then Supreme Council and the Executive Board. Later on its name was changed as Improved Seed Enterprise (ISE) and 21 farms were transferred for seed production of various crops. These farms had a total area of 11768.27 ha and 5504.87 ha area under cultivation. The Headquarters of the Enterprise was in Kabul and it had several sub-offices in different provinces. The Improved Seed Enterprise and the Department of Extension and Production started improved seed production of vegetables through the UNDP/FAO assisted Seed Production and Training Project, AFG/86/010 (Phase I and II). The project established various facilities such as seed testing laboratories, vegetable seed processing facilities, five small seed production and training farms and a training center. Great emphasis was placed on improving trained manpower in various aspects of seed production technology. The activities of this project were later taken over by Seed Improvement Programme of Northern Afghanistan (AFG/93/001) with the main objective to promote the production and multiplication of high quality cereal seeds and seedlings of vegetables and saplings of fruit trees for distribution to farmers with a view to increase supply of food, fresh vegetables and fruits necessary for improving the nutritional standard of the people.

The above arrangement continued up to April 1992. At the time of collapse of Dr. Najibulah's government, all facilities were lost due to large scale looting which followed the change in the regime and continued for several years. The linkage and coordination between central office and provincial offices was completely broken down and few provincial units are now working directly under the control of respective provincial authorities. Four provincial units of Improved Seed Enterprise namely Herat, Pulekhumri, Mazar and Kabul are still functioning as Implementing Partners (IPs) of the seed programme. The breeder and foundation seed are produced at different farms of ISE while quality seed is produced under the contract seed producer's scheme with the assistance from seed component of AFG/96/004. The sale proceeds of breeder and foundation seed is used by the individual ISE units to pay salary and wages of its staff and to meet day to day expenditure. It has created an extra enthusiasm to produce more seed so that more could be earned. Until further institutional support can be developed, these arrangements are likely to continue.

Following the signing of the Geneva Accord in 1988, a second focal point based in Pakistan was opened for across-border areas not controlled by government. The first seed orders were placed with seed growers in Pakistan during 1988. The base was set up in Islamabad. Sub-bases are in Peshawar (for access by the Khyber and through Khost) and Quetta (for access by the Khojk pass). From then until the conclusion of the campaign December 1997, more than 48,413 MT of high quality seeds of various crops have been delivered to more than 1,000,000 rural families (approximately 7.50 million people) in Afghanistan, in almost all provinces. Up to 1995 the programme was funded through the United Nations Office for the Coordination of Humanitarian and Economic Assistance Programmes for Afghanistan (UNOCA). Later major contribution came from a core fund of UNDP. A number of international donors namely the Governments of the Netherlands, Japan, Canada, Switzerland, Norway and the EEC also contributed generously. The life of the present programme is up to December 1999.

1.1. Objectives

The main objective of the Seed Programme has been to establish basic household food security in accessible areas of Afghanistan and, in the process, to develop an in-country sustainable capacity for seed multiplication to internationally accepted standards, in support of Afghanistan's agriculture rehabilitation and development process.

The Afghanistan Seed Programme could be divided into two phases namely: Emergency phase (1988-1994) and Rehabilitation phase (1995 onward). Accordingly, objectives of both phases slightly differ from each other.

Emergency seed supply programme

The project was part of the operation Salam and from the outset, operation Salam and its UN Family Members recognized the importance of improved seed as a catalyst for agricultural, relief, rehabilitation, and development. Under FAO leadership, the team approach to emergency seed aid was further developed, and ultimately included in the UN Plan of Action, 1990. One apparent consequence of the breakdown of the agricultural system has been a degeneration in the genetic potential of principal field crops due to the lack of timely replacement with new seeds. The initial aim of the FAO programme was to get as much good seed of suitable varieties to as many farmers in the region as possible, depending on availability of resources. The Action Plan envisaged the following objective:

"The major thrust of the seed programme will be the reintroduction of improved seed, since there has been a drastic genetic deterioration of seed materials during war years. The supply of good certified seed of suitable varieties and in sufficient quantities will be assured. In parallel, assistance will be provided to strengthen the institutional arrangements for the production of improved seeds."

The emergency seed programme was based on the following premises:

- to support returnees, and

- to restore food production through a package of inputs.

To encourage displaced populations and refugees to return to formally productive agricultural area, two requirements were paramount:

to help rehabilitate neglected and abandoned irrigation systems, because agriculture in these regions is particularly dependent on irrigation; and to ensure that, as far as possible, returnees and resident farmers have access to good quality seed of high yielding, disease resistant varieties of crops suitable for local conditions along with matching quantity of fertilizer.

Rehabilitation phase

The main objective was to establish basic household food security in accessible areas of Afghanistan and, in the process, to develop an in-country sustainable capacity for seed multiplication to internationally accepted standards, in support of Afghanistan's agriculture rehabilitation and development process. This includes:

- expanding the in-country seed production and processing capabilities to meet farmers demand of various seeds;
- expanding the seed production and processing capabilities to meet national demands;
- establishing a contract seed distribution network in various provinces;
- promoting national enterprises to enter into seed business;
- assuring, through the provision of high quality seed (supplied with matching fertilizer), timely, in-country production of seed for major crops; and
- establishing a network of seed units in the areas not covered earlier, specializing in maintaining, securing and producing seed of various classes through appropriate implementing partners (NGOs).

1.2. Implementation Methodologies

Emergency phase

The Geneva Accord was signed in February 1988, and humanitarian and economic assistance programmes mounted soon after. Operation Salam was officially launched on October 12, 1988 coordinated by Prince Sadruddin Aga Khan. Following the signing of the Geneva Accord various aid agencies including Food and Agriculture Organization of the United Nations (FAO) launched emergency seed programme in Afghanistan. It was clear at the outset that significant improvements in crop production were required to feed the resident and returnee population of the country and to reduce the necessity for food aid. Seed was one of the essential tools to achieve this goal, and was therefore a major priority in all agricultural rehabilitation programmes.

The first funds for the FAO seed programme based in Pakistan became available to the FAO through UNOCA during September 1988, resulting from a donation from the Government of the Netherlands. Follow up funds were also provided through UNOCA, by the Government of Japan, for various aspects of the Agricultural Rehabilitation Programme, of which the seed programme was part.

Immediately scanning of all available research and trial records was started to identify suitable cultivars so that seed of these could be procured from the country of origin. The cultivars were identified on the basis of: (a) previous trials conducted both inside Afghanistan, and other agro-ecological analogues, as well as, (b) international advice from Afghan specialists, FAO Experts, and the appropriate CGIAR Centers. Fortunately, there is considerable agro-ecological overlap between Afghanistan and other neighboring countries. Historically this has facilitated the systematic identification and import of seeds of new plant varieties. This made the task of identification of adapted varieties easy.

Naturally, in the beginning sufficient quantity of certified seed of "identified" crops and cultivars was not available any where in the world. Accordingly, a systematic "seed industry approach" had to be developed. Because of the agricultural calendar, this involved ordering seed, up to eleven months ahead of delivery. Considering the short period between harvesting and planting seed of many these crops and cultivars in the Northern Hemisphere, it was decided to order fresh seed from neighboring countries such as Pakistan, India and Iran so that it could be delivered to staging areas for Afghanistan well in time for onward dispatch, according to the local agricultural calendar. Peshawar and Quetta were selected as staging areas due to their proximity to Afghanistan and presence of more than three million refugees in these towns. "Recommended" wheat cultivars Pak-81 and PS-85 and Sonalika and HD-2285 for example, were developed in Pakistan and India respectively by leading wheat breeders, from CIMMYT material. Their participation, and that of concerned seed certification agencies considerably reduced the risk, and cost of developing a separate emergency seed infrastructure for producing early generation and certified seed in support of the Afghan seed programme.

From the beginning of the programme, it was known that there are numerous other national selections available in neighboring countries from material resulting from the CIMMYT wheat programme which also had/have a very good chance of performing as well as, if not better than identified varieties. Search for other varieties continued and small quantities were procured and tested inside Afghanistan. During 1988-89, a total of 2,300 MT of certified wheat seed was ordered/delivered to staging areas close to the Pak/Afghan border (Quetta and Peshawar) according to international standards and prices.

Considering that funds will be available for next four -five years orders for seed of selected crop species following recognized international seed industry standards and procedures were raised in the field as either Local Purchase Orders, or Purchase Requisitions. All seed was purchased according to international standards, at international prices, on the basis of guidelines cleared through FAO HQ, based on ISTA recommendations. These guidelines were, by design, suited to rehabilitating the seed industry in Afghanistan, as well as providing emergency seed aid.

In Pakistan, seed was produced on order both by the public and private sectors with the strict supervision of Federal Seed Certification Department (FSCD) under Pakistan's Seed Act, including the services of concerned:

(a) Officers of the Technical Division, FAO HQ;

(b) The consultants and trained Afghan graduate consultants; and

(c) Qualified international and national scientists and technicians, including concerned plant breeders from Pakistan Agriculture Research Council (PARC) and National Agriculture Research Council (NARC).

This seed was produced over and above the local targets to avoid any inconvenience to the host country and negative effect on its seed supply plan.

All seed was treated with fungicide and insecticide. The contents of bags were clearly marked on bags including warnings (in English and Persian) of the toxicity of seed to humans and animals. Seed was supplied double bagged, complete with a certificate of origin (including pedigree), FSCD quality certificates, a valid phytosanitary certificate, and independent superintendence, arranged by FAO HQ.

Seed deliveries were made as a package with fertilizer. Suitable national and international NGOs were appointed as Implementing Partners (IPs) to distribute seed and fertilizer inside Afghanistan. All seed was distributed on the basis of formal project documents, including a project proposal, letters of agreement and shipping papers. Back up support, training, and planning for the future were integral parts of all sub projects signed with IPs. World Food Programme (WFP) moved the seed and fertilizer both inside Pakistan and across border, under formal agreements with FAO and NGOs respectively.

Within the first two years a well-defined system of seed production, procurement and distribution was specified and continued till throughout the emergency phase. Seed for the next year across border programme was ordered during the fall of previous year. The programme of the subsequent year included several additional varieties that showed promise inside Afghanistan. As well as producing certified seed this phase of the programme included an important element of "early generation" seed starting from progeny rows through breeder/ nucleus and pre-basic, to basic seed to ensure the independent continuity of the programme and provide the basis of a rehabilitation programme in Afghanistan, as soon as the situation allowed.

Rehabilitation phase

Ensuring food security in the entire country in general and in chronically food deficit areas in particular became a paramount consideration of the implementation strategy. It is well known that seed security and food security comes one after another in the sequence of an agricultural rehabilitation programme. Accordingly, seed security and food self-sufficiency became a central theme of all activities.

The Afghanistan Rehabilitation Strategy - Action plan for Immediate Rehabilitation recommended that high priority might be given to increase the in country seed production to meet farmers demand and to increase food production. For this purpose, the EU provided more than US $ five million for the project-Seed Supply and In-country Seed Production Programme in Afghanistan. Though funds were available but it was not an easy task to produce seed inside Afghanistan. Physical and administrative infrastructures are prerequisite to produce seed. There was no problem with regard to physical facilities. Afghanistan has dry and cool climate which is considered ideal for quality seed production and Afghan farmers had sufficient land, water and labor to grow their own seeds. Therefore, necessary physical facilities were available for initiating an in-country seed production programme. However, administrative infrastructure was badly hit by years of war, the central administrative and technical arms of government have suffered high levels of physical destruction and staff losses during the internecine fight in the country. Governmental institutions, having lost nearly all vehicles and equipment, and fraught with staff uncertainties, officials have understandable difficulties in delivering the level of technical assistance and physical support to deal with the problems effectively with which they are confronted. The strict enforcement of the policy devised by various commanders hardly a formal operative government system exists. Besides, the social processes and local customs were required to be taken into consideration when initiating the in-country seed programme. The security of staff and equipment was another major problem. Frequent evacuations and suspension of activities were common.

Under such a situation it was very difficult to adhere to a specific time schedule while timeliness of all operations is an important aspect of seed production. If even a single operation is missed, the quality or quantity, if not both, of seed may suffer. A vigorous and flexible calendar of operations for all stages had to be drawn up in advance and vigorously implemented. The experience gained by project staff and IPs in the emergency phase proved very useful. The implementation strategy was devised on the basis of experience gained in the past and prevailing situation in Afghanistan.

Such complex problems called for an imaginative implementation methodology of diversification and sustainability to remedy the situation rather than pursuing a typical programme. Therefore, an unorthodox approach was followed. Instead of depending on one organization to produce seed, four different institutions were used. These include:

- Parastatal organization: Provincial units of ISE to multiply breeder and foundation seed under direct supervision of the project.
- State farms: To produce Quality declared seed and testing of new varieties.

- NGOs: Five well-established international and one national NGOs were given production targets of quality declared seed under contractual obligations. Wide spread presence of these NGOs helped to cover more area under seed production and distribution.

- Community: In PEACE Initiative districts seed was produced directly with communities. Under this scheme, communities were asked to form seed groups to produce and distribute seed directly under supervision of the project.

Due to food insecurity local commanders did not permit movement of seed from their region and it was next to impossible to move seed from an area controlled by rival commander. Therefore, economic principle which dictates production of seed in a compact area was deliberately ignored to reach more farmers in isolated and inaccessible areas. Therefore, decentralization of the seed production was preferred. The centralized approach envisages concentration of the seed production, processing and storage in the most favorable areas from the point of view of agro-climatic conditions, high yields, and low prices, and the provision of the technical services of the highest order with large-sized modern processing plants. However, in contrast a diffuse approach consisting of large number of small seed producing pockets was adopted to ensure that farmers of every region and every stratum of society could benefit from the programme.

The problems did not end with the successful production of seed in a decentralized manner. The procurement of seed from seed growers posed a serious problem. In a country where sense of insecurity is always present, purchasing seed against cash was difficult. Besides, hyperinflation made it difficult to fix prices for a given time. Seed growers refused to hand over the seed against cash payment on earlier agreed procurement price due to the fear of food shortage. To overcome this problem, a scheme known as Food-for-Seed was launched with the collaboration of the Word Food Programme (WFP). Under the scheme, seed growers were given food wheat at 1:1.25 seed - food ratio. The seed so procured was processed, packed and sold at reasonable price. The sales price was decided in consultation with Agency Coordinating Body for Afghan Relief (ACBAR). The sale proceeds was kept in a separate account and its further use is required to be made only for strengthening of seed activities. More than US $ two million have accumulated as sale proceeds till date.

Considering that substantial funds will be generated, transparency and efficient management of funds were essential. Accordingly, a Seed Review Group (SRG) was constituted to review the progress of the project twice a year. The SRG meets every six months and includes representatives of all the UN, NGO and Donors Organizations with a stake in the seed programme in Afghanistan. Similarly, a Fund Management Committee (FMU) composed of seed producers, FAO, WFP and UNDP was constituted and use of funds was authorized by the committee after evaluation of each proposal submitted by IPs. IPs are required to maintain a separate account to deposit sale proceeds and furnish account in each SRG meeting. All NGOs have their accounts in Pakistan thus there was no problem of banking but ISE, a parastatal organization is based in Afghanistan where no proper banking facilities exist and hyper inflation is integral part of economy, has problem to manage the funds. Efforts were made to invest the funds to create and improve seed facilities but still it could not use total sale proceeds. To avoid the diversion of funds to war machine by local authorities, an ISE fund management account was opened in Islamabad and unutilized funds were converted into dollars and deposited in this account. So far, more than US $ 120,000 have accumulated in this account. The funds could be used to rehabilitate seed facilities as soon as a proper government becomes a reality.

The polycentric approach, food-for-seed, sale of seed and fertilizer, generation of funds and its deposit outside the country and several other non conventional steps taken to meet challenges posed by natural or man made disasters ensured a fail safe mechanisms needed in a stalled system. It also increased cost effectiveness and efficiency in the delivery system.

2. EXPECTED VERSUS ACTUAL RESULTS

Expected	Actual
Emergency Phase.	
To provide emergency support to returnee farmers for their agricultural re-settlement in Afghanistan and to restore agriculture for increased food production through the provision of a package of agricultural inputs (seed, fertilizer, farm equipment, etc.)	By 1995, more than 27,372 MT of winter crops seeds and 6,571 MT of summer crop seeds (including 2,400 MT sugarcane) had been distributed (Table 1). In addition, fertilizer (Table 2) and root-stock, cuttings, and saplings of commercially important trees (Table 3) were also distributed.
Rehabilitation Phase.	
The project was designed to procure 1,500 MT seed internationally and produce 3,000 MT seed inside the country. Thus, distribution of 4,500 MT seed of various crops was envisaged. Besides, required quantity of foundation and breeder seeds was to be produced. The following targets for seed and fertilizer were fixed for 1995-96 and 1996-97.	In comparison to targets the project could succeed to procure and produce more than 8,665.50 MT seed, 192% more than the target (Appendix I). Similarly, against procurement of 3,500 MT fertilizer, the project procured 5,318 MT i.e., 152% more than the target (Appendix-II). The project imported only less than 100 MT seed and balance quantity was produced inside the country itself.
SEED	**SEED**
Crop Target (MT)	Crop Achievement (MT)
Wheat 4,000	Wheat 5,895
Barley --	Barley 271
Rice 100	Rice 696
Maize 350	Maize 1,135
Pulses 150	Pulses 579
FERTILIZER	**FERTILIZER**
DAP 1,500	DAP 4,118
Urea 2,000	Urea 1,200

2.1. Were expected results achieved?

No other report can provide an answer better than the evaluation reports prepared by various external evaluation missions fielded by UNDP and FAO. The main conclusions relating to seed are quoted below:

Quote:

(i) In spite of the poor security situation, project Seed Production and Training (AFG/86/010) was able to accomplish its objectives. <u>In fact, it has exceeded its set targets</u>.

(ii) For project Programme for Rehabilitation of Afghanistan Agriculture (AFG/92/008) the procurement of seed has had a major positive impact in Southern/Eastern regions. The efforts of the project have been monumental, in view of the difficulties encountered in the process of seed distribution, a fact which is recognized and appreciated by the farmers and NGOs involved in the process.

(iii) Project Seed Improvement Programme of Northern Afghanistan (AFG/93/001) has implemented activities in new areas as well as devise the application of new programme such as:

- New sites in Northern and Western Afghanistan
- Farmers to Farmers Seed Exchange Programme
- Creation of funds for counterpart management from the sales proceeds of seeds and fertilizers. (1994)

Integrated Crop and Food Production in Afghanistan (AFG/94/002)

Seed

(i) This component is making excellent progress. The quality of seed crops varied from good to excellent with likely yields ranging from 3-5 MT/ha. (1996)

Food Security through Sustainable Crop Production (AFG/96/004)

Seed

(ii) The component has set and maintains high technical standards and its results are impressive by any measure. The seed production system is technically sound and commercially viable. (1998)

Unquote

The seed/fertilizer delivery and distribution programme has been a remarkably successful one. Approximately, 1,000,000 Afghan farm families have received **<u>direct</u>** inputs of high quality seed plus fertilizer since the programme started in the autumn of 1989. The incremental benefits were much more than the production, procurement and distribution of quality seed and fertilizer. The seed distributed under the project has serviced the area earlier sown under old and traditional varieties. Yield in these areas, is assumed to have declined considerably over the years of constant use of seeds of old and degenerated varieties. A minimum yield of 2-3 MT/ha has been achieved by the shift from traditional seed (low technology) to new improved seed (high technology) farming in comparison to existing average productivity of 0.8- 1.0 MT/ha. The high technology is based on a package of inputs of which seed is but one. It is difficult to determine empirically the proportion of the yield increments due to each element of the input package distributed under the project. The assumption used here, that 50 percent of the yield impact is due to seed and fertilizer supplied under the project, is believed conservative. By bringing large area under improved seed created opportunity for seasonal employment. Though it is difficult to translate the benefits exactly into dollars and incremental food production in MT and it could be a matter of discussion but certainly production and distribution of quality seeds and fertilizer has made significant impact on crop yield and food security in target areas. The seed was used nationwide.

Improved seed supplies and reduced prices have markedly improved small farmers access to essential inputs.

The achievements should not be measured only in terms of the crop yields being obtained and arable land abandoned during the war, brought back into production, but also in terms of successfully delivery and distribution of these inputs in a country that has been and continues to be in a state of civil war.

2.2 Why the expected results achieved

The important aspects for achieving desired results are explained below:

Resilience

Shock and pessimism are natural-aftermath of any disaster or catastrophe. However, one of the most important things to remember when planning for the revival and rehabilitation of agricultural production systems after a major catastrophic or destructive event, is that unless the systems had exceeded their human support capacities, and if the basic resources of land and water have not been affected, the human and agro-ecological systems are likely to be highly resilient. This is particularly true if the production system was previously robust. Actual recovery rates very often outstrip desk-top projections, even more so if strategic inputs are supplied. This has been true with respect to the seed programme under study.

The agricultural production systems of Afghanistan can only be described as robust and resilient. For twenty years, from 1978 to 1998, rural production systems in Afghanistan have continued to support the remaining rural population under conditions of extreme difficulty. Although malnutrition and hunger were reported, this did not degenerate into a situation as catastrophic as in the African countries of Somalia, Mauritania or Ethiopia for example.

Unorthodox approach

The country has long been known for its independence, for its aggressive and xenophobic warrior farmers and pastoralist, for rulers renowned in court cunning and cruelty, and for an ability to maintain a unique balance between independence and outside influence and dominance (Toynbee, 1969). Afghanistan, therefore, is considered a most difficult place to work with. Nothing is quite as straightforward as it may seem in Afghanistan, it is a very complex situation. Such a complex problem called for an imaginative seed programme of diversification, sustainability and collaboration with other UN agencies to remedy the situation rather than pursuing a typical seed development programme. Therefore, an unorthodox approach was followed. Instead of depending on one organization to produce seed, different institutions were used. Inputs were sold. Non-conventional procedures were followed to manage the funds and direct contact and confrontation with local commanders was avoided.

Decentralization

To reach more farmers in isolated and inaccessible areas, decentralization of the seed production was preferred.

Food-for-seed

A new scheme known as food-for-seed in collaboration with the World Food Programme (WFP) was launched. The in-country seed production oriented programme requires tapping of local resources supported by new techniques. Accordingly, thousands of seed growers were registered to produce seed throughout the country. Hyperinflation made the purchase of raw seed against cash problematic and, therefore, Food-for-seed scheme was developed. Contract seed producers received food wheat at a 1.25: 1 exchange ratio for the seed they handed over (there are variations for other crops). This ensured that seed growers had enough food to feed their own families and that a large proportion of their seed crops would travel out of the production areas to other farmers specially in food deficit areas. It also helped to make considerable savings of funds. Food-for seed began in 1994 with 300 MT; 2,607 MT in 1995; 6,905 MT in 1996 and 6,578 MT in 1997.

Mobility

The contract seed producers were scattered over a wide area in various provinces. It was highly unlikely that seed growers would take their produce long distance for threshing, processing and treating, therefore, threshers and seed cleaners needed to be mobile. More than 15 mobile seed cleaners were used for processing seed.

NGOs

Before 1978, there were few NGOs working inside Afghanistan. However, following the war several international organizations came to help the Afghans. NGOs operate in all provinces of Afghanistan and mostly based in rural areas. The Seed Programme in emergency phase has been a mosaic of hundreds of NGO-run micro-projects, ranging from a five-ton seed scheme to a Jerib of fruit trees nursery. At present, one national and five international NGOs are assisting the programme to meet its seed production and distribution targets. The contribution of NGOs has been very vital and the programme could not have done what it has done without them. All NGOs working with Seed Programme have been advised to privatize the seed business and survive on their own, either as commercial development organization or individual entrepreneurs.

Risk

Working in Afghanistan is full of risk. No body knows how many Kalashnikovs are in the country. A traveler is always on a hair-trigger. When one enters the field, he struggles against 'sleeping enemies' left behind the war which stand ready to maim a trespasser whether he is a seed technologist or a farmer. More than 10 million mines exist in the country. Radical local customs introduced recently have added another dimension to the list of potential danger. FAO had been unusually lucky in this respect, no lives were lost due to shooting or mines. However, project staff (an international staff member along with national staff) were once taken hostage for several days. Such risks were always present. The national and international staff was psychologically prepared for these risks and therefore, able to achieve the targets.

3. CONCLUSIONS AND RECOMMENDATIONS BASED ON LESSONS LEARNED

The decision to include a Seed Programme as one of the short-term emergency relief programmes launched under Salam Operation by UN has proved beyond any doubt that seed and fertilizer were as important to returnee and resident farmers as any other inputs such as cash, ration, shelter material and medicine. It has also confirmed that seed availability and food security comes one after another in the sequence of an agricultural emergency/rehabilitation programme.

Successful implementation has also demonstrated that in spite of a series of unsatisfactory security situations, serious risk to the lives and property of the staff of the project, irreparable physical damage to the various facilities, a complete blockade of certain provinces, time to time temporary suspension of activities due to evacuation or other reasons, difficulties in importing supplies and equipments and their transfer across the border, most project activities were completed on schedule and a very good base has been established for a viable in-country seed production and distribution programme in accessible areas of the country. No where in the world had such as a large quantity of specially selected, high quality seed and fertilizer been delivered so successfully in a country still in the throes of a civil war.

The most significant lesson of all that FAO has learned in Afghanistan is the importance of trusting the individual's own thrust for development. By keeping its intervention small and sharply focused on specific areas and problems it has been able to provide the initial lift that the majority of Afghan farmers are in need of before embarking on their own climb upwards to self-sufficiency. In the present case, the programme could not distribute seed and fertilizer equally in all provinces, rather the difference was from 10 MT to a high of 7000 MT, but still the effect on crop productivity was quite significant at national level.

Emergency seed supplies are made in response to a food crisis. In such a situation the speed of response is critical, and blanket distribution of commercial/certified seed is the only practical solution. However, any provision of such seed must not only be well selected and appropriate to the environment into which it is being sent, it must also be of the highest possible quality. This is essential if agriculture is to get off to a flying start, to delay the process of degeneration as long as possible and protection against exotic pests and diseases.

It was also realized that only effective way to operate the seed programme of this kind on the ground is for the field implementation to be carried out by Afghans.

Interventions and inputs must be kept to the most essential and to a level of technology well within knowledge, experience and skills of the rural population. Introduction of hybrids and varieties excessively dependent on high doses of fertilizer, pesticides and insecticides should be avoided.

After farmers receive improved seed (together with other associated inputs) as emergency assistance, their requirements quickly progress from the emergency, to rehabilitation and into a development mode. Often this occurs even before the end of the first production cycle. Farmers themselves do not differentiate between these various phases of emergency relief, rehabilitation and development. On this basis, the programme also treats them as a continuum under its integrated strategy.

FAO's Afghanistan seed programme provides many useful lessons that might be adopted and/or adapted to other situations elsewhere in the world.

*****

Table 1: Seed distribution (tonnes) under the FAO programme, 1988/89 to 1994/95

Winter crops					
Season	**Wheat**	**Barley**	**Peas**	**Vegetables**	**Berseem**
1988-89	2665	-	-	-	-
1989/90	3825	-	-	0.228	-
1990/91	2408	-	-	0.975	-
1991/92	3888	8.0	-	0.635	-
1992/93	5141	8.0	-	0.475	-
1993/94	3622	9.7	-	2.000	1.70
1994/95	5780	8.6	0.5	0.700	1.85
Total	**27329**	**34.3**	**0.5**	**5.013**	**3.55**

Summer crops						
Season	Rice	Maize	Green gram	Ground nut	Sugarcane	Sesame
1988-89	-	43	-	-	-	-
1989/90 -		233	-	-	-	-
1990/91	25.5	284	-	-	-	-
1991/92	36.2	916	10	-	-	-
1992/93	38.0	581	17	2	800	-
1993/94	50.5	311	34	-	800	0.5
1994/95	51.5	1505	30	3	800	-
Total	**201.7**	**3873**	**91**	**5**	**2400**	**0.5**

Table 2: Distribution of diammonium phosphate (DAP) and Urea fertilizer, 1989/90 to 1994/95

Season	**DAP (t)**	**Urea (t)**
1989/90	-	5500
1990/91	-	-
1991/92	1853	1500
1992/93	3229	-
1993/94	2000	2000
1994/95	2050	3550
Total	**9132**	**12550**

Table 3: Distribution of saplings, apple root stock and popular cuttings, 1989/90 to 1994/95

Season	Saplings (1)	Apple root stock	Poplar (*Populus nigra*) cuttings
1989/90-		-	-
1990/91	207 770	16 510	120 000
1991/92	414 650	10 000	-
1992/93	359 650	6000	10 000
1993/94	309 600	11 000	40 000
1994/95	52 800	-	50 000
Total	**1344 470**	**43 510**	**220 000**

(1). Saplings of apple, pear, apricot, peach, cherry, almond, and persimmon trees

Case Study

EMERGENCY PRODUCTION OF HIGH-QUALITY SEED FOR WAR AFFECTED AREAS IN BOSNIA AND HERZEGOVINA

FAO/TCOR - UNDP project financed by the Government of Japan (BiH/96/028)

Mona S. Chaya
FAO Emergency Coordinator for Bosnia and Herzegovina

1. OVERVIEW OF PROJECT ACTIVITIES

1.1 Background

As a result of the war, the local seed industry in Bosnia and Herzegovina (BiH) was broken down. Seed production systems have undergone serious damages which will take several years to repair and rebuild prior capacities. Seed quality control and phyto-sanitary capacities were completely devastated where equipment were totally destroyed or stolen, and qualified staff had left. During and after the war, seed production was being carried out in a completely chaotic way without following the minimum standards of quality seed production. Seed certification was totally by-passed due to lack of seed quality testing facilities and an absence of official inspection. No control on seed import was made which allowed the entrance into the country of huge amounts of new and unadapted varieties not on the official list and sometimes introducing new diseases due to lack of quarantine measures. Normally, a new variety is tested for two or three years in experimental plots before being accepted or rejected by the country. In many cases, distributed humanitarian seed aid, especially wheat seed, were not quality tested and were of poor quality. All these factors led to extremely low yields and a complex seed production situation that should have been seriously treated on the ministerial level.

1.2 Objectives

The FAO project BiH/96/028 "Emergency production of high quality seed for war affected areas in BiH" was initiated in BiH under very chaotic conditions in the seed industry with the initial objective of assisting its rehabilitation.

The project's immediate objectives are the re-establishment of an early generation seed production, maintenance and multiplication programme of major agricultural crops in BiH and the establishment of a seed quality and phytosanitary control capacity.

The overall objectives are to: (i) reduce dependence on direct food aid and external sources of agricultural assistance, especially seed; (ii) assist the country to become more self-sufficient in seed production in terms of quality and availability and to produce surplus for sale; (iii) re-establish marketing systems in order to improve the sustainability of local production; and (iv) encourage the emergence of localized seed supply networks.

1.3 Implementation methodologies

The project started in October 1996 for a duration of 18 months (October 1996 - May 1998) which is exactly 1 ½ growing seasons. The crops dealt with were wheat and potato and to a lesser extent bean, onion and alfalfa (*Medicago sativa*). The operations were carried out on a ratio basis of ¾ in the entity Federation of BiH (FBiH) and ¼ in the entity Republika Srpska (RS) for political reasons. According to the Dayton Agreement peace accord, signed in November 1995, the two entities FBiH and RS constitute BiH.

The official implementing partner is the Ministry of Agriculture, Water Resources and Forestry (MAWRAF). In FBIH, MAWRAF designated the Agricultural Institute for Development, UPI, as implementing partner and in RS, the other MAWRAF designated Banja Luka Agricultural Research Institute as the implementing partner for wheat and Sokolac Potato institute for potato.

All project operations were made after discussions between FAO experts and the implementing partners. Decisions on varieties, quantities of required seed, categories of seed, locations of planting, time of planting and target beneficiaries were jointly made between FAO and the implementing partner.

It is worth mentioning that the project was strictly dealing, in all five crops, with high quality seed or (primary seed) which are the highest categories in the seed multiplication chain and these are Super Elite (SE), Elite (E) and Original (O). When the project started, it was the only one dealing with such kind of seed. All other humanitarian seed distributions were of the commercial category. Due to the delicacy of handling primary seed, and due to their high value, the main target beneficiaries were seed producers capable of handling them. Seed which reached the commercial stage in the seed multiplication cycle, were distributed to around 2000 families of vulnerable farmers.

The project involved 60 producers, and around 1300 ha of planted area in 25 locations, all over BiH in its both entities. It was the first project of its kind which was implemented in an interactive manner between the two entities of BiH where seed was purchased from one side and planted in the other. This way, the peace objective of the Dayton accord was enhanced.

1.4 Expected results

According to the project document the expected results were:

a. Depending on weather conditions, the expected quantity of high quality seed produced at the end of the project should be up to 10 000 mt potato, 1 200 mt cereals, 500 mt onion (set), and 100 mt beans including a proportion used to continue multiplication of super-elite and elite seed for a total estimated value of US$ 11 000 000.

b. Establishment of an operational seed quality control and phytosanitary laboratory.

2. EXPECTED VERSUS ACTUAL RESULTS.

2.1 Were expected results achieved?

The expected results were achieved to a large extent. The second objective which is establishing an operational seed quality control and phytosanitary laboratory was achieved to a higher level than expected. The laboratory rebuilt and equipped under the project is testing seed according to international standards under the guidance of an FAO expert who is training the national technicians. The laboratory is testing and issuing valid certificates to seed produced all over the country which will enable BiH to soon export seed. The laboratory will soon be accredited by the International Seed Testing Agency which is the reference international agency for seed quality control. This part of the project has taken so much importance that it became a separate project with an independent budget.

As for the first objective of the project to produce specific amounts of seed, it was fully achieved for wheat, in terms of quantity and quality and achieved to a lesser extent in quantity but fully in quality for potato, bean, onion and alfalfa. Therefore, we can say that qualitatively, the project fully succeeded for this part and quantitatively to a lesser extent.

2.1 If so, what where the enabling conditions ?

The need for all activities of the project was great. This was of the main reason for success. The project was fully supported by the Ministry of Agriculture and the designated implementing partners, who saw its benefits.

Briefly, the enabling conditions could be, *inter alia,* the following:

a. The project was designed after an assessment mission which studied the needs of the country in terms of seed industry rehabilitation.
b. The project has been designed as an emergency activity bridged towards development. It started right after the end of hostilities with a long term objective. This is why the impact of the project has been so wide and was well appreciated by all collaborators.
c. All activities of the project were executed and decisions made after thorough and careful discussions with the national implementing partners and the ministries of agriculture. Several agricultural institutes were consulted for the choice of varieties and locations of planting. Trust in the project was gained from the nationals which made advice given by FAO experts of value to them.
d. Rapid action/quick response from FAO HQs for approval or disapproval on requests from the field rendered decision making quicker, allowing quicker action.
e. Coordination with other agencies dealing with seed distribution prevented overlap of activities, and FAO's advice was well respected and sought.

2.2 If expected results were not achieved, why not?

a. Climatic conditions were not very favorable.
b. Problems with quality at the start of the project were more acute than what FAO consultants thought. This is why production figures improved during the second multiplication cycle after the establishment of the quality control laboratory.

CONCLUSIONS AND RECOMMENDATIONS BASED ON LESSONS LEARNED

3.1 Conclusions

In spite of all the difficulties and constraints encountered by the project, it is considered as successful and has achieved its objectives to a large extent and in some cases more than expected. The project had a big impact on the rehabilitation of seed industry in the country and is considered as the only serious and organized attempt to rehabilitate it following international standards and imposing quality measures. In spite of its emergency nature, the project had a developmental objective which made the impact more anchored.

In addition to the ministries of agriculture, the project involved many national agricultural institutes, local producers, local suppliers, etc., and coordinated with other international agencies and NGOs dealing with seed distribution. This led to less mistakes and liability in decision making.

3.2 Recommendations

The following recommendation can be made:

a. Close consultation with the national authorities is very important. Besides the ministries of agriculture, the opinion of agricultural research institutes proved to be quite valuable for the decisions on technical aspects.
b. A thorough assessment of the situation prior to designing the project or writing the project document is quite important. Once the consultants or executers are in the field, and before making final decisions, a quick re-assessment of the situation in terms of needs (quantities and varieties), climatic conditions, other agencies's involvement, market information etc… would be very useful.
c. Rapidity in action: this is what characterizes emergency operations even if they were bridged towards development. Any attempt to improve emergency operations towards more sustainability should not ignore this characteristic.
d. The consultants or experts should be aware of the political situation in the country.
e. Technically sound agricultural experts, with knowledge of different disciplines of agriculture would be very useful and would prevent many mistakes. Motivation and conscientiousness are important factors as well.

Case Studies

LES DISTRIBUTION GRATUITES D'INTRANTS AGRICOLES ET LES PROGRAMMES DE MULTIPLICATION DE SEMENCES AU RWANDA DE 1994 A 1998

Jean-François Gascon
FAO Emergency Coordinator
Kigali, Rwanda

Abstract

Soon after the 1994 incident, United Nations Agencies, humanitarian NGOs, the bilateral and multilateral donors, combined efforts to mobilize emergent assistance to Rwanda. The country was then faced with a disastrous socio-economic situation. In the field of agriculture, most of the farmers have been forced to left their home for almost a long period and the main priority needs of those who have already been resettle were seeds and agricultural tools. However, other emergent needs were equally identified: collection, transportation and storage facilities for locally obtained seeds, protection against livestock diseases, production, processing and distribution of seeds and, at last, plant protection.

As a whole, from1994 to 1998 (eight agricultural seasons), emergent needs were estimated at 70 million US $, out of which 46 million was obtain. The funds mobilized could allow purchasing 38,000 tons of seeds, transporting and distributing them to Rwandan farmers. This was an average of 3 kg of seeds per household per season. The greatest part of seeds distributed were beans (64% of the total quantity). During the 8 agricultural seasons, 4,270,000 hoes were freely given to farmers, an equivalent of 3 hoes per household.

The funds obtained were, however, relatively limited. Limited in ratio to the farmers self-provision (a Rwandan household usually uses an average of 20 kg of seeds per season). Limited in relation to the farmers' qualitative needs: the biggest part of funds mobilized was allocated to seed distribution and a very little part was used in seed multiplication. Limited in relation with quantitative needs: only 66% of the required seeds and tools was obtained. And, at last, limited as in comparison to resources given to other sectors – food aid particularly (as for Rwanda, the funds mobilized for the agricultural sector, in 1996 for example, were equivalent to the amount expended in only 4 days in Rwandan refugees camps outside the country).

In September 1994, due to multitude and diversity of partners, the Ministry of Agriculture, with FAO assistance, introduced a coordination mechanism for emergent agricultural operations. This coordination was comparatively effective and managed to harmonize different approaches use, to organize interventions, to establish a quality control mechanism and minimize risks in seeds distribution.

For the distribution of seeds, it was advice to focus on local varieties or on those widely spread in neighboring countries with ecological similarities. As for beans, an important staple food crop in Rwanda, the distribution of mixed local varieties, as much as diversification is possible, was emphasizing.

As regard seed production, since the 1994 incident, the FAO is aware of the serious qualitative and quantitative shortage of seeds in the country. Unfortunately, the sector of seed multiplication has so far received very little support from donors and very few organizations are involved in this sector. On its side, the FAO managed 3 projects.

The major aim of these projects was to enable farmers' associations to produce commercial seeds approved by the Selected Seeds Department. Contracts for the production of these seeds were signed between MINAGRI, FAO and these associations. These contracts were signed in such a way that the FAO would supply them with necessary agricultural inputs – seeds, fertilizers, rhyzobium and pesticides – and buy the produced seeds at a price prior fixed; while the MINAGRI would realize the sanitary and technical controls. Seed production was accompanied with training courses. A guidebook on the production of quality commercial seeds in Rwanda was published and distributed by the FAO.

A boost in the agricultural production achieved in the country since 1994 is observed. This positive trend is due to a number of factors. Although, the role of seeds, tools and fertilizers distribution contributed to a small extent to this achievement. Among the four main groups of food crops – cereals, pulses, banana and roots and tubers – assistance was mainly given to cereal and pulses production. These productions which had tremendously fallen in 1994 has regained its former production level. So, the food shortage decreases and is now estimated, for the second semester of 1998, at 63,000 tones cereal-equivalent.

In the last few years, production and distribution of seeds in Rwanda were faced with the following constraints :

- *Prior 1994, no crucial programme was established to increase commercial seed production. There existed some efforts but monopolized by the public sector and managed by agricultural support projects. After 1994, therefore, it not only meant to rehabilitate a system but to imagine a new concept and to put in place a new system, something that made matters much more complicated;*
- *Once funds are mobilized the state of emergence dictates immediate effects. However, on national and international markets, vegetable seeds are the only ones with acceptable quality. Other types of seeds proposed by regional suppliers were grains primarily meant for food; and once cleaned could be considered as seeds. Besides, it was too late to carry out any seeds production tasks according to financial backers' timetable. The only alternative was the purchase of what was available both at national and regional markets, as well as emphasizing on quality control;*
- *In respect to seed production, the main problem was to mobilize funds and to get some basic seeds. The quantities offered by the Selected Seeds Department were relatively low;*
- *Further, lack of Plants' Protection Department particularly worsen the stock control and management; and*
- *The absence of seeds legislation laws and lack of national programme for seeds production restrain both the public and private organizations full involvement in the sector.*

In conclusion, the following recommendations can be drawn :

- *In case of disaster, seed importation from neighboring regions becomes imperative because of the absence, before this disaster, of a guiding policy, law as well as national seed production programmes. It is recommended that the above should be done, not during the time of disaster but before. It is important to note that the MINAGRI is now finalizing seed legislation and a national programme for seed production. Once the textbook is published, it shall work as a guideline for future interventions;*

- *Purchasing grains directly from local markets is advantageous, because it facilitates quick supply of seeds, but disadvantageous in terms of quality. The donors' approval to postpone seed distribution for a period of six months and produce instead quality seeds must be explored;*
- *After a serious disaster, disaster relief actors are numerous and the risk of committing mistakes also become high. In this manner, an effective coordination is necessary. That is to say, better understanding of the agricultural situation, better technical assistance, better sanitary controls and better follow-up;*
- *In accordance with the Agricultural Development Strategy laid by the MINAGRI, the position and role of various partners are well prescribed. The ISAR,SSS, the Agricultural Department, farmers associations and agricultural support institutions, each of the above has a limited role on his own, which can not be over stepped, even in times of disaster; and*
- *An important lesson to keep in mind is that, despite a high number of emergent humanitarian partners, a negligible number concentrate in agriculture. In addition, it is important to note that farmers' financial ability and their own investment are often largely under-estimated, whereas external partners' expectations are over-estimated. Such poor superficial evaluation by external partners can lead to big mistakes, mischief and disillusion.*

1. VUE D'ENSEMBLE SUR LES ACTIVITÉS DES PROGRAMMES ET PROJETS

1.1. Objectifs

Immédiatement après les événements de 1994, les Agences des Nations Unies, les ONG humanitaires et les bailleurs de fonds bi- ou multi-latéraux se mobilisèrent rapidement pour une assistance d'urgence au Rwanda. Le pays était alors confronté à une situation économique et sociale désastreuse avec plus d'un million de victimes du génocide, de formidables mouvements de population, près de deux millions de réfugiés à l'extérieur et le retour de plusieurs centaines de milliers de personnes exilées dans les pays voisins depuis 20 ou 30 ans. Sur le plan agricole, la plus grande partie des exploitations avaient été abandonnées durant une période plus ou moins longue et les premiers besoins exprimés par ceux qui se réinstallaient étaient les semences et l'outillage agricole. D'autres besoins d'urgence furent également rapidement identifiés: collecte, transport et stockage des semences locales, protection sanitaire du bétail, production, conditionnement et distribution des semences et protection des végétaux. Cependant, la réponse des bailleurs de fonds concerna presque exclusivement les distributions gratuites de semences et d'outils agricoles. Comme le montre le graphique ci-après, l'exemple des financements mobilisés par la FAO, de 1994 à 1998 au Rwanda, est à cet égard significatif: sur les 14 millions de $ reçus pour les opérations d'urgence, 44 % furent destinés aux distributions de semences, 24 % pour l'outillage agricole, 10 % pour les fertilisants, 6 % seulement pour les programmes de multiplication de semences, 5% pour les distributions d'animaux, 2 % pour le reboisement et 2 % pour la protection du bétail.

L'objectif des distributions était de répondre rapidement aux besoins des agriculteurs en semences et en outils afin de les aider à reprendre leurs activités de production. Pour les deux saisons agricoles qui suivaient les événements, l'ensemble des ménages réinstallés devaient être assistés. Par la suite les distributions furent ciblées sur les ménages en difficulté.

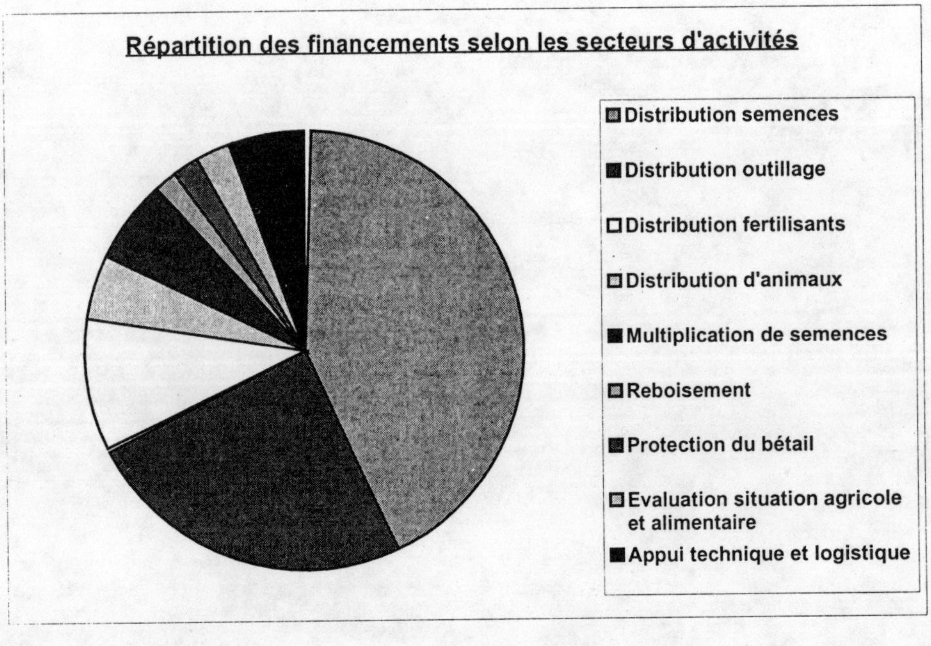

1.2. Méthodes et procédures de mise en place

En juillet 1994, la FAO, dans le cadre de l'Appel Consolidé Inter-Agences et sur la base d'une première appréciation des besoins, demanda une assistance en semences, houes, fertilisants et produits phytosanitaires pour 400.000 ménages. En septembre, le Ministère de l'Agriculture, de l'Elevage, de l'Environnement et du Développement Rural (MINAGRI) mettait en place, avec l'assistance de la FAO, une Coordination des opérations agricoles d'urgence. Cette Coordination qui regroupait tous les intervenants du secteur agricole, précisa les besoins sur la base du nombre de ménages qui se réinstallaient effectivement et des quantités moyennes de semences utilisées habituellement par chaque ménage.

Pour les saisons qui suivirent, les calculs de besoins furent conduit par cette Coordination à partir du nombre de ménages en difficulté selon les 154 communes du Rwanda, des superficies moyennes cultivées selon les productions et les préfectures et des quantités moyennes de semences utilisées par unité de surface selon les cultures.

Pour les distributions, la recommandation générale était de se limiter aux variétés utilisées localement ou, à la limite, aux variétés répandues dans les régions écologiquement similaires des pays voisins. Pour le haricot, culture de grande importance alimentaire au Rwanda, il fut impérativement demandé de distribuer des mélanges, les plus diversifiés possible, de variétés locales à l'instar des habitudes paysannes qui veulent que les haricots nains soient toujours semés sous la forme d'un mélange de nombreuses variétés (plus de 10 généralement).

Pour la production de semences, notons que dés la fin des événements de 1994, la FAO, consciente du problème de la disponibilité et surtout de la qualité des semences, rechercha des financements pour apporter un appui à la production de "semences commerciales" (autrement qualifiées de "semences de qualité déclarée"). Le premier niveau de production de semences, à savoir les "semences de souche", est, au Rwanda, du ressort de l'Institut des Sciences Agronomiques du Rwanda (ISAR) qui est assisté par les instituts internationaux de recherche. Le deuxième niveau, qui est la production de "semences de base" est du ressort du Service des Semences Sélectionnées (SSS) au MINAGRI qui reçoit un appui de la Coopération Belge. Jusqu'à présent, le troisième niveau – production de "semences de qualité déclarée" – n'a bénéficié que d'un appui très limité et n'a fait l'objet que de quelques interventions ponctuelles des ONG du secteur agricole. De son côté, la FAO a pu conduire 3 projets visant, entre autres activités, à produire des semences de qualité déclarée.

Ces projets ont eu pour objet de faire produire des semences commerciales par des groupements ou associations paysannes à partir des semences de base acquises auprès du SSS. Des contrats de production ont ainsi été établi entre ces associations, la FAO et le MINAGRI. Ces contrats prévoient la fourniture, par la FAO de tous les intrants nécessaires (semences, engrais, rhyzobium, pesticides), l'achat des semences produites à un prix fixé dés le début de la campagne et un contrôle sanitaire et technique de la part du MINAGRI. Les semences produites sont destinées à approvisionner des associations paysannes en charge de personnes en difficulté. Les semences sont remises gratuitement à ces associations qui les revendent à leurs membres. Ces derniers bénéficiant alors d'un crédit remboursable en espèces ou en nature.

1.3. Résultats attendus

Le résultat final attendu de ces opérations de distributions ou de production de semences était de limiter le déficit alimentaire en relancant la production vivrière et, pour ce faire de fournir aux agriculteurs les intrants les plus demandés.

Les besoins en intrants ont évolué avec les saisons en fonction de la situation agricole du moment et des mouvements de retour des populations qui s'installaient ou se réinstallaient dans les exploitations agricoles.

Evolution de la population résidante rwandaise de 1994 à 1998

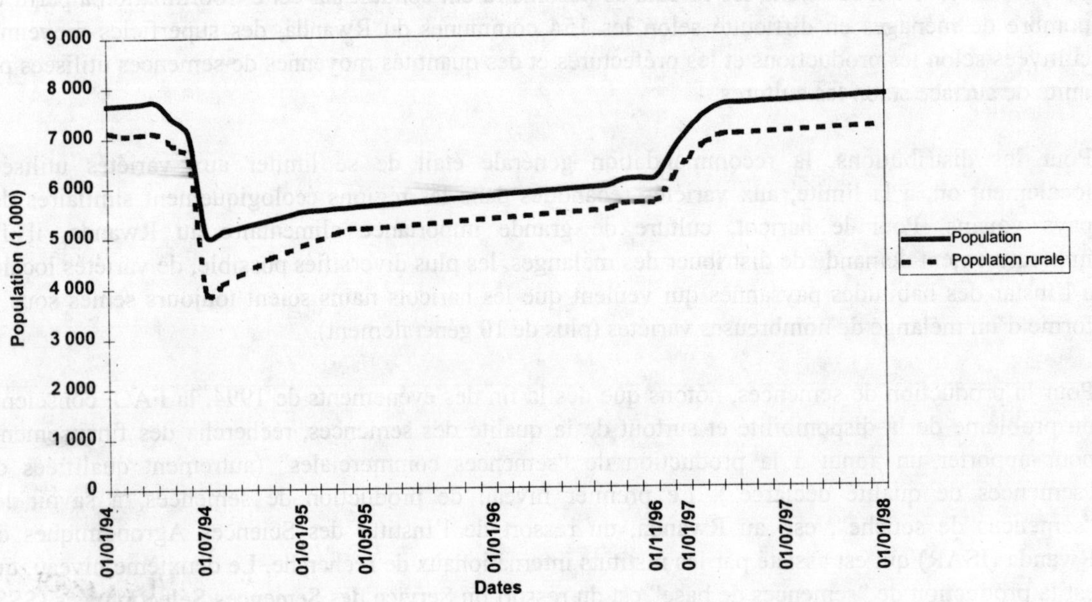

2. LES RÉSULTATS PAR RAPPORT AUX PRÉVISIONS

2.1. Les résultats attendus ont-ils été atteints ?

Ces résultats peuvent se mesurer aux niveaux de la quantité de semences produites, de la quantité d'intrants fournis, de l'évolution de la production agricole et de l'évolution du déficit alimentaire.

2.1.1. Quantités de semences produites

La dispersion et le caractère ponctuel des interventions font qu'aucune statistique n'est actuellement disponible sur la quantité globale de semences produites au Rwanda entre 1994 et 1998. Sur les trois projets conduit par la FAO, un est achevé, il s'agit du projet financé, en 1995-96, par la Banque Mondiale et qui comprenait un volet de production de semences. Les deux autres projets, financés par la Belgique et la Suisse, sont en cours d'exécution. Seuls les résultats du premier projet sont indiqués ici :

- Saison 1995 B: production de 22,2 tonnes de semences de haricot (variétés H75, A321, PVA 779, Umubani, Flor de Mayo);

- Saison 1996 A: production de 46,92 tonnes de semences de haricot (variétés Vuninkingi, Umubano, PVA 779, RWK 5, RWK 719, Ngwinurare, Flor de Mayo);
- Saisons 1995 B et 1996 A: production de 1.518.000 boutures de manioc (variétés Murindi, Kiryumukwe, Cleorina et Eala);
- Saison 1995 B et 1996 A: production de 8.600.000 boutures de patate douce (variété Mugande).

La production de ces semences a été accompagnée de sessions de formation à l'intention des responsables des associations partenaires et des techniciens du MINAGRI. En outre, grâce à ce projet, un manuel ***"La production de semences de qualité déclarée au Rwanda"*** a été produit et diffusé par la FAO.

2.1.2. Quantités d'intrants distribués

Pour ce qui est des intrants fournis nous noterons que la mobilisation des fonds pour les opérations d'urgence a connu son apogée quelques mois après le génocide pour dépasser, en saison 1995 B, les 13 millions de $. Par la suite, les distributions se ralentirent pour n'atteindre, en 1997 A, qu'un million de $ environ. Les retours massifs et médiatiques de novembre 1996 relancèrent les distributions qui, en 1997 B, s'élevèrent à plus de 5 millions de $. La tendance fut ensuite à la baisse et, en 1998 B, les intrants agricoles distribués gratuitement représentaient environ 3 millions de $.

Au total, entre 1994 et 1998 (huit saisons agricoles), pour des besoins estimés à 70 millions de $, 46 millions de $, soit 66 % de ces besoins, furent mobilisés.

Ces fonds permirent l'achat, le transport jusque dans les communes et la distribution de 38.000 tonnes de semences soit une moyenne de 3 kg de semences/ménage/saison. Cette assistance est relativement modeste car, chaque saison, un ménage utilise, en moyenne, une vingtaine de kg de semences. La plus grande partie des semences distribuées était constituée de haricots qui, à eux seuls représentaient 64 % du tonnage. Au cours des 8 saisons agricoles, 4.270.000 houes furent également remises gratuitement aux agriculteurs, soit, en moyenne, presque 3 houes par ménage.

Les principaux fournisseurs furent: l'Union Européenne (Programme d'appui à la sécurité alimentaire au Rwanda), la FAO, le HCR, l'UNICEF, l'USAID, la GTZ, le CICR et diverses ONG parmi lesquelles, World Vision, Care International, FHI ou ZOA. Les contrôles de qualité (humidité, puretés spécifique et variétale, taux de germination) furent réalisés par le laboratoire du Service des Semences Sélectionnées au MINAGRI. Les transports jusque dans les communes et les distributions furent d'abord de la responsabilité des ONG et organisations diverses opérant sur le terrain puis dirigées par les autorités préfectorales et communales.

Montants mobilisés pour les distributions gratuites d'intrants agricoles, par l'ensemble des intervenants ($ EU)

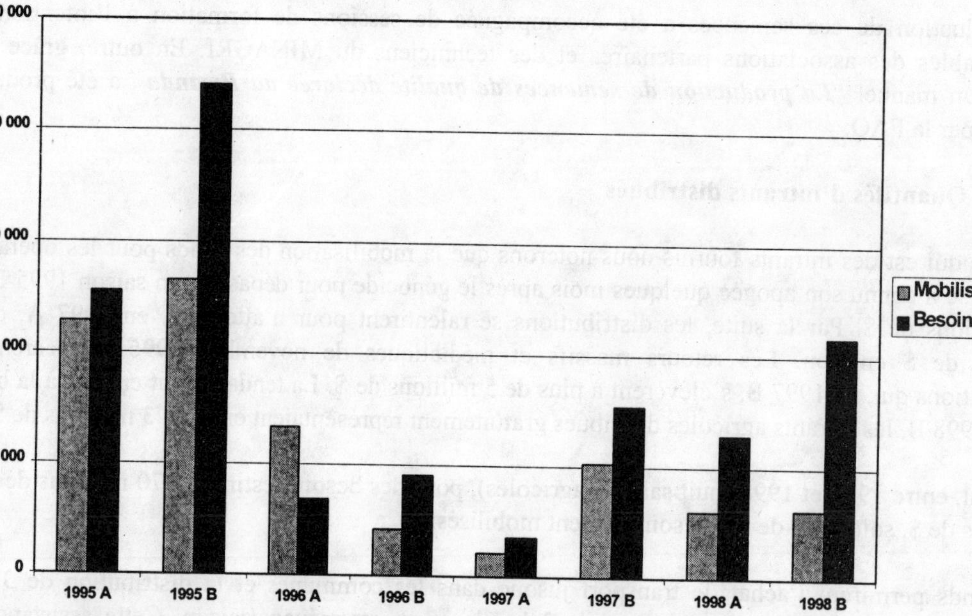

Quantités totales de semences distribuées gratuitement de 1994 à 1998 (tonnes)

Désignation	Quantités	%
Semences de haricot	24 386	64%
Semences maraîchères	173	0,5%
Semences de maïs	4 613	12%
Semences de soja	2 035	5%
Semences de petit pois	1 155	3%
Semences d'arachide	172	0,5%
Semences de sorgho	2 300	6%
Semences de blé	736	2%
Semences de pomme de terre	2 335	6%
Semences de riz	35	0,09%
Total	37 940	100%

Répartition des semences distribuées

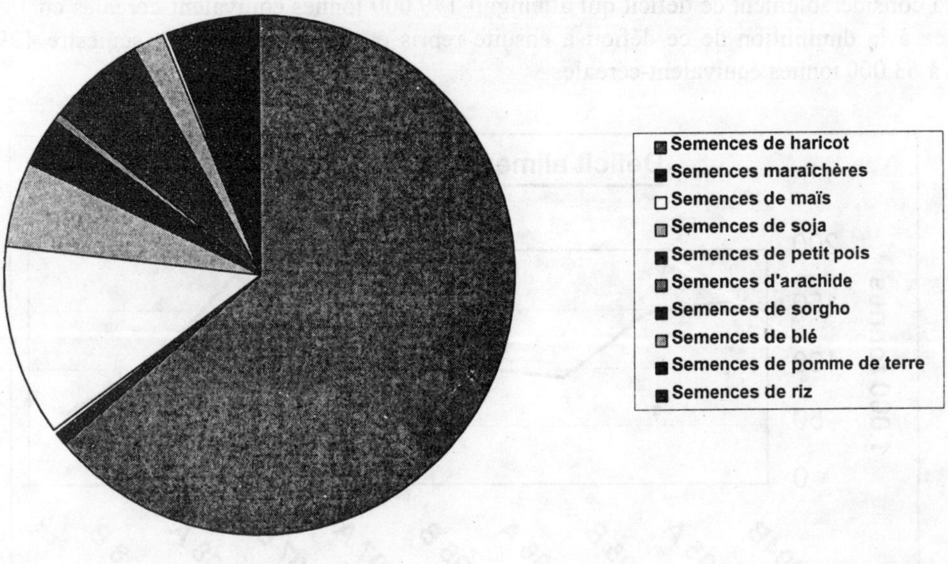

2.1.3. La relance de la production agricole

Cette relance est due à de nombreux facteurs et les distributions de semences, d'outils agricoles et de fertilisants n'en constituent qu'une fraction très modeste. Il est en effet clairement établi que, d'une part, les agriculteurs disposent de ressources notables mais presque toujours sous-estimées et que, d'autre part, les distributions ne représentent finalement qu'une part très réduite des semences et intrants utilisés. Parmi les quatre grands groupes de productions vivrières – céréales, légumineuses, bananes et racines & tubercules – les appuis en semences/boutures ont quasi-exlusivement porté sur les céréales et les légumineuses. La production des légumineuses qui avait particulièrement chuté en 1994 a très vite retrouvé sa place. Les mauvaises conditions climatiques de 1997 ont à nouveau fait chuter les productions qui ont repris leur progression en 1998. Les céréales ont connu également une bonne relance mais qui s'est ralentie en 1998 à cause de problèmes climatiques.

Evolution des productions

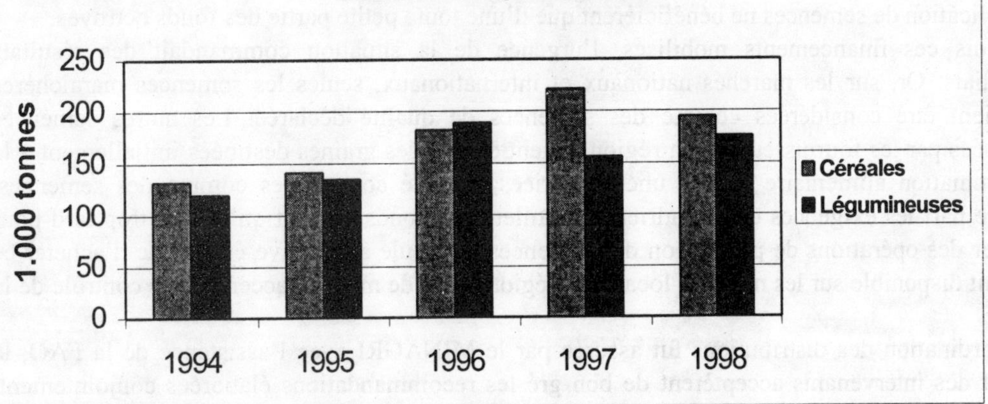

2.1.4. L'évolution du déficit vivrier

Entre septembre 1994 et novembre 1996, le déficit alimentaire s'est progressivement réduit grâce à la relance de la production agricole. En novembre 1996, le retour de plus de 1,4 millions de personnes aggrava considérablement ce déficit qui atteingnit 179.000 tonnes équivalent-céréales en 1997 B. La tendance à la diminution de ce déficit a ensuite repris et, pour le deuxième semestre 1998, il est estimé à 63.000 tonnes équivalent-céréales.

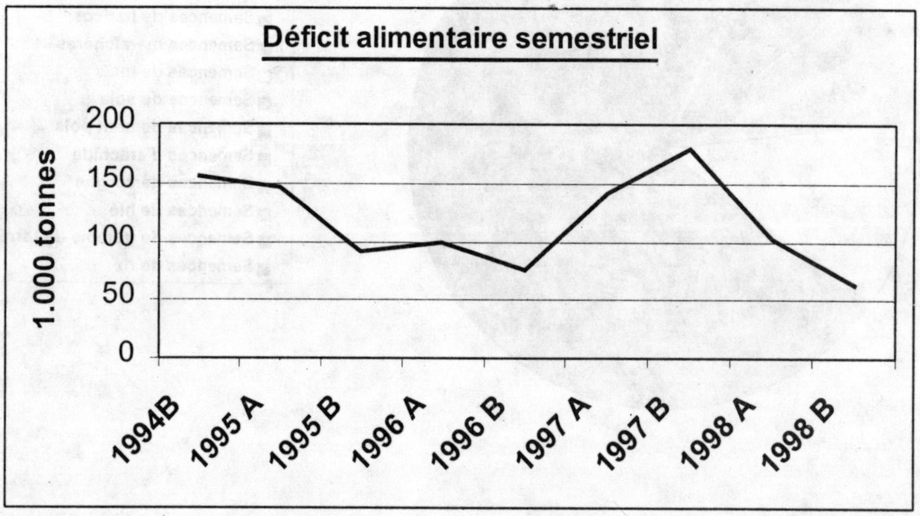

2.2. Les contraintes

Le principal objectif des opérations agricoles d'urgence était de fournir, en temps voulu, des semences et de l'outillage en qualité et en quantité suffisante pour permettre aux ménages en difficulté de relancer leurs activités de production. Par rapport à cet objectif nous remarquerons que:

- Avant 1994, aucun programme important de multiplication de semences commerciales n'avait été mis en place. Seules quelques actions ponctuelles et monoplisées par le secteur public avaient été conduites dans le cadre de projet assistant les Services Agricoles. Le résultat est, qu'après 1994, il ne s'agissait pas de réhabiliter simplement un système mais de concevoir et de mettre en place une nouvelle organisation, ce qui est déjà beaucoup plus compliqué.
- Suite aux appels de fonds, les financements mobilisés ne couvrirent qu'en partie les besoins (66%) et les opérations de distribution reçurent la plus grande part. Les opérations de multiplication de semences ne bénéficièrent que d'une toute petite partie des fonds octroyés.
- Une fois ces financements mobilisés, l'urgence de la situation commandait des résultats immédiats. Or, sur les marchés nationaux et internationaux, seules les semences maraîchères pouvaient être considérées comme des semences de qualité déclarée. Les autres semences proposées par les fournisseurs de la région étaient en fait des graines destinées initiallement à la consommation alimentaire et qui, une fois triées, ont été considérées comme des semences. D'autre part les exigences de calendrier des bailleurs de fonds ont fait qu'il était trop tard pour engager des opérations de production des semences. La seule alternative était donc d'acheter ce qui était disponible sur les marchés locaux ou régionaux et de mettre l'accent sur le contrôle de la qualité.
- La coordination des distributions fut assurée par le MINAGRI avec l'assistance de la FAO, la plupart des intervenants acceptèrent de bon gré les recommandations élaborées conjointement.

Néanmoins, certaines organisations passerent outre et organisérent des distributions contraires à ces recommandations. C'est ainsi que des variétés non inscrites dans le catalogue du MINAGRI furent distribuées, que du matériel importé se révéla porteur de maladies (boutures de manioc importées par SOH et porteuses du virus de la mosaïque) ou encore que d'importantes quantités de haricot ne comptant qu'une seule variété (K20 ou Coco Rose) furent importées d'Ouganda (Union Européenne spécialement) en contradiction à la consigne de rechercher des mélanges les plus riches possibles.

- Pour les opérations de production de semences, la principale difficulté fut de trouver les semences de base nécessaires. Les capacités du SSS de fournir ce type de semences sans délai étant relativement limitées.
- L'absence d'un Service de Protection des Végétaux compliqua singulièrement les contrôles et les traitements des stocks.
- L'absence jusqu'à ce jour d'une législation semencière et d'un Programme National de production de semences entrave l'engagement des organisations publiques ou privées dans le secteur semencier.

3. CONCLUSIONS ET RECOMMANDATIONS

❖ L'inexistence d'une politique, d'une législation et d'un Programme semenciers font, qu'en cas de crise, le seul recours est l'importation de semences à partir de zones écologiquement proches. Ce n'est pas pendant une crise ou immédiatement après qu'il est possible de concevoir et d'implanter rapidement un système, a priori lourd même si une période d'après-crise est parfois le moment opportun pour lancer de nouvelles opérations. Par contre, dans le cadre d'interventions destinées à prévenir ou à limiter les effets de catastrophes, ce type d'action devrait être systématiquement envisagé. Notons qu'à l'heure actuelle, le MINAGRI finalise une législation semencière et un programme national de production de semences. Ces textes, une fois publiés, guideront les interventions futures.

❖ Les achats directs de graines sur les marchès locaux présentent l'avantage de la rapidité mais l'inconvénient de la qualité. L'acceptation par les bailleurs de fonds de "décaler" les distributions d'une saison (6 mois) en produisant d'abord les semences voulues mérite d'être recherchée.

❖ Après une crise, les intervenants sont nombreux et de multiples dérapages sont observés. Une coordination effective, c'est-à-dire maîtrisant bien la situation agricole du lieu, capable d'apporter un appui technique solide, d'imposer les consignes élaborée d'un conjointement, de contrôler les importations et d'assurer un suivi phytosanitaire, est indispensable pour limiter les dérives.

❖ Dans la Stratégie de Développement Agricole élaborée par la MINAGRI, la place et le rôle des différents intervenants sont bien précisés. L'ISAR, le SSS, les Services Agricoles, les opérateurs privés, les organisations paysannes et les institutions d'appui ont chacun un mandat qu'il convient, même dans les situations difficiles de ne pas outrepasser.

❖ Une leçon importante qu'il convient finalement de bien retenir est que, malgré le caractère massif des interventions d'urgence humanitaire, la part consacrée à l'agriculture est extrêmement limitée. Dans le cas du Rwanda, l'aide apportée au secteur agricole sur un an (1996 par exemple) équivalait au montant dépensé en seulement 4 jours dans les camps de réfugiés rwandais situés à l'extérieur du pays. D'autre part, il convient d'être bien conscient que les appuis apportés aux agriculteurs ne représentent qu'une très petite fraction de leur investissement total. Leurs capacités financières et leur investissement personnel sont, le plus souvent, largement sous-

estimés et, a contrario, les effets attendus des interventions extérieures sont quant à eux généralement sur-estimés. Cette mauvaise appréciation initiale du poids respectif des intervenants peut entraîner de graves erreurs quant au choix des appuis à apporter et conduire à tort, en fin d'opération, à une appréciation désenchantée des interventions extérieures.

Case Studies

THE EFFECTS OF THE RWANDAN WAR ON CROP PRODUCTION, SEED SECURITY AND VARIETAL SECURITY: A COMPARISON OF TWO CROPS[1]

Louise Sperling
Senior Scientist
International Center for Tropical Agriculture (CIAT)
The Hague, The Netherlands

1. INTRODUCTION

The escalation of the Rwandan civil war in April 1994 resulted in the death of up to one million persons and the displacement of another two million. Agriculture, the main occupation of upwards of 90% of the population, was acutely affected as civil disruptions peaked in the middle of a major growing season. Overall harvest losses during this period were officially estimated to be as high as 60% (Dr. Iyameremye n.d).

The aid community feared the worst in terms of farmers' ability to find their agricultural bearings again after the war. Non-governmental organisations (NGOs), United Nations agencies, and bilateral donors responded swiftly and on a wide scale to the post-genocide crisis, with some 30 different agencies intervening in emergency seed provision during the first season alone. Massive amounts of bean, maize and vegetable seed poured in, along with widespread distribution of hoes, fertilizer, and pitchforks (MINAGRI/UNREO/PNUD/FAO, 1994).

"Seeds of Hope" (SOH)[2], an unconventional aid intervener, similarly responded along its own areas of expertise. Among its key activities (Table 1), this coalition of national and international

---

1. A slightly modified version of this article was published in the Overseas Development Institute's Agricultural Research & Network series (AgREN, Network Paper No. 75, July 1997), with Diana Carney offering superb editorial advice. An addendum (section 7) has also been added for this FAO-sponsored Workshop on 'Developing Institutional Agreements and Capacity to Assist Farmers in Disaster Situations to Restore Agricultural Systems and Seed Security Activities."

This assessment work was generously supported by IDRC (the International Development Research Centre). Special thanks are extended to Joachim Voss and Luis Navarro. Colleagues at ISAR, CIP, and CIAT also provided logistical support and important research insights to this work on beans and potatoes: the substantial help of Steve Beebe, Robin Buruchara, Peter Ewell, Ngerero Nkuriza, Neeroj Pant, Marie-Jean Uwera and Wayne Younquist was very much appreciated. This paper is largely drawn from SOH Assessment Documents 1 to 10.

2. The Seeds of Hope Initiative was formalised in September 1995. Many African NARS (those of Burundi, Ethiopia, Kenya, Malawi, Tanzania, Uganda, Zaire and Zimbabwe) contributed germplasm, field space, and advice to the initiative. In addition, some eight International Agricultural Research Centres were heavily involved in the Rwandan Agricultural reconstruction, namely:

   International Centre for Tropical Agriculture (CIAT)
   International Maize and Wheat Improvement Centre (CIMMYT)
   International Potato Centre (CIP and its network PRAPACE)
   International Centre for Research in Agroforestry (ICRAF)
   International Crops Research Institute for the Semi-Arid Tropics (ICRISAT)

agricultural research centers promoted intensive varietal and genetic assessments, and rapid multiplication of landraces and select improved varieties for possible reintroduction into stressed agricultural systems. SOH felt its skills were complementary to NGO/UN strengths--and much needed in a country so genetically-remarkable as Rwanda: both beans (*Phaseolus vulgaris, L.*) and sorghum (*Sorghum bicolor*) were notable for their wide pre-war genetic variability. While, in one sense, the genetic heritage of Rwanda was of concern to the world community: pre-civil war, for example, farmers grew the greatest range of bean varieties found in active use anywhere, genetic diversity was also deemed integral for rebuilding and sustaining productive agricultural systems for Rwanda's farmers.

This article focuses on seed and varietal assessments and describes some of the precise effects of war on Rwandan agricultural systems. This task is made easier by the perhaps unique amount of immediate post-war fieldwork and pre-war research which was conducted in Rwanda. During the first three post-war seasons, SOH undertook surveys and intensive interviews at both a regional and nationwide level, in order to plan complementary work to that of seed relief agencies. SOH looked at farmers' management of beans, potatoes, sorghum, maize and cassava. This paper, however, focuses primarily on two crops – beans and potatoes – in order to draw out insights into the *differential* nature of crop survival during times of acute stress.

In terms of pre-war research, the paper draws on work done on the Rwandan bean and potato systems over a period of 10 years or more. The research was conducted in the context of a regional bean network, RESAPAC,[3] (which received technical support from CIAT, the International Centre for Tropical Agriculture) and a regional potato network, PRAPACE[4] (which received support from CIP, the International Potato Centre). Research had been conducted in areas such as: varietal improvement; food processing and storage; soil erosion and fertility enhancement; and seed multiplication options.

An introductory caveat is necessary. At the time of writing, the Rwandan situation remained fluid; guerilla attacks continued and refugee camps still housed Rwandans outside their country's borders. The findings and projections of this piece refer to the situation as of November 1996. At that time, over one million Rwandans (approximately one eighth of the total population) remained in neighbouring Zaire, Tanzania and Burundi. The return of several hundred thousand of these refugees in December 1996 may have altered the bean and potato seed need profiles, although the author expects that the varietal situation, and hence diversity assessments, remained unchanged.

International Institute of Tropical Agriculture (IITA)
International Livestock Research Institute (ILRI)
International Plant Genetic Resources Institute (IPGRI)

3. RESAPAC = Réseau pour l'Amélioration du Haricot (*Phaseolus vulgaris*) dans la Région de l'Afrique Centrale.

4. PRAPACE= Programmeme Régional d'Amélioration de la Culture de la Pomme de Terre et de la Patate Douce en Afrique Centrale et de l'Est.

Table 1. Overview of Activities of the Seeds of Hope

EMERGENCY PERIOD
* Assist NGOs in sourcing relief seed
* Multiply landraces and key improved varietal materials for possible reintroduction
* Effect varietal and genetic inventories to compare pre- and post-war crop diversity
* Assess impact of seed aid
REHABILITATION PERIOD
* Restore germplasm collections and active breeding material to NARS
* Help to restore formal seed multiplication capacity
* Train new cadre of national scientists in on- station and on-farm research
* Assist in re-establishing scientific facilities

2. BACKGROUND: RWANDAN AGRICULTURE

Though a tiny country (26,000 km^2, traversable by car in just three hours), Rwanda harbours enormous ecological diversity. The countryside is punctuated by hundreds of rolling hills, giving rise to myriad social and agricultural micro-niches. Altitudes range from 1,000 to 4,500 metres and rainfall varies from about 800 to 1,600 mm/annum.

Landlocked and without important mineral deposits, Rwanda relies on the increasingly stressed agricultural sector as the backbone and blood of its national economy. Agriculture in pre-war Rwanda exhibited all the characteristics and problems of small farm agriculture. Large families cultivated garden-sized plots with nothing but hoes and limited amounts of manure. Fields were intensively intercropped (bananas, beans, local greens, sweet potato, cassava) and with an exploding population (3.7% per annum), farms of less than a hectare were increasingly fragmented into 5, 10, or even 20 separate parcels.

National indicators suggest how stretched this central African country was becoming, even before the war. Population density was the highest on the continent (445 persons/km^2 arable land), per capita income among the lowest ($US 285 p.a.) and chronic malnutrition touched over 50% of children under the age of six. Omnipresent poverty appeared to be exacerbating what was already a tradition of highly individualistic (and stressed) social ties. Hunger leads to desperation. For instance, in the early 1990s, pre-war, it was not too unusual to hear of someone being beaten to death by neighbours for stealing some cassava roots. Histórically, such serious communal sanctions were purportedly reserved only for theft of high-value livestock.

It would be inaccurate to speak of any one crop as *the* key to Rwandan agriculture. Bananas, used mainly for making beer, are the lynch pin of many social transactions and are a good income generator. Beans are the 'meat' of the countryside. Cassava and sweet potatoes are what really keep poor stomachs full and sorghum is vital for making children's porridge. This article focuses on beans and potatoes because of the interesting contrasts they offer in terms of seed needs and varietal diversity, rather than for any reasons of overall importance in Rwandan agricultural systems.

3. OVERVIEW OF BEAN AND POTATO GENETIC DIVERSITY AND PRODUCTION: PRE-WAR

3.1 Beans

Pre-war, beans were a relatively high value crop grown by over 90% of farmers in all regions of Rwanda. In the years immediately preceding the civil war, Rwandan bean consumption per capita was the highest in the world (50 kg p.a.). The rich could afford to eat beans twice or three times a day, while the poor would sell part of their bean stocks to purchase lower value cassava (the money generated from the sale of 1 kg of beans buys 4 kg of starch – the difference between filling four stomachs versus eight).

Although they originated in MesoAmerica and the Andean region, beans (*Phaseolus vulgaris, L.*) have developed an important secondary centre of diversity in the central Africa region. It is thought that they were brought to eastern Africa by Portuguese traders in the 16th century. They seem to have made their way to Rwanda between 1576–1609, when *mwami* (king) Kigeli II sent forces to quell opposition in part of the present-day area of Cyangugu, a prefecture in western Rwanda.

Most Rwandan farmers grow beans during both major cropping seasons, with a good number trying to intensify production for a third season in the lowland valley bottoms. Farmers tend to grow the same varieties for both home use and sale. Beans are grown in mixtures of anywhere from 3–30 components, with a single farmer sometimes managing two or three different varietal blends.(Lamb and Hardman, 1985; Voss, 1992). Unusually, Rwanda's growing urban population has not yet demanded that the rural producers 'homogenise' their product to a single uniform variety; mixes of varieties still sell at roughly the same price as pure strains.

It is important to note that free bean seed exchange was not very common in pre-war Rwanda. Best friends, immediate family, or dear neighbours might get a handful of seeds (say 100g) here and there, but this would certainly not be enough to make a dent in sowing needs. Substantial quantities of seed were therefore routinely obtained through off-farm informal channels, mostly purchased from local markets, or sometimes from neighbours or town merchants. Studies suggest that the formal sector only supplied about 2–3% of the bean seed sown in pre-war Rwanda, although this seemingly token amount was of great importance in that it was the main source of new varieties.

A number of studies suggest how vast pre-war bean genetic diversity in Rwanda may have been. The *Institut des Sciences Agronomiques du Rwanda* (ISAR) collections were said to have contained some 700 local varieties and a USAID/ISAR study conducted in 1984 identified 284 major components in major bean-producing zones (Lamb and Hardman, 1985). In 1992 researchers collected 5,000 farmer mixtures from all regions of the country except for the north (where the first incursions of the war had already begun) and isolated 545 different grain types (Scheidegger, 1993).

While the number of bean phenotypes in active use is impressive, their spatial distribution is also notable. Relatively few varieties are found throughout the country, partially because bean production zones are so diverse, extending from 1,000-2,200 meters. However, even within zones, over short distances, bean variability is great. For example, a 1991 study in southern Rwanda sampled 75 mixtures collected within 10 km radius and found that over 30% of the mixtures were grown in only a single farmer's field (Cishahayo *et al.*, n.d.).

3.2 Potato

Like beans, potatoes (*Solanum tuberosum*) originate in the Americas, in this case in the Andes. However, they are a much more recent introduction to Rwanda, having arrived during the period of German colonisation (1894-1916). Their nomenclature illustrates these origins. The first Rwandan epithet for potatoes, '*intofanyi*', is a corruption of the german '*kartofen*', while the current name, '*ibirayi*' dates from World War II when the Belgians took over the role of colonisers. '*Ibirayi*' derives from '*uburayi*', meaning 'that which is from Europe'.

Potatoes play a very different role to beans in the lives of Rwandan farmers. They are cultivated on a much more limited scale, mostly at higher altitudes (above 1,900 m), although potato production has recently been increasing in the mid-altitudes (particularly in marginal areas of the South) as an important poverty reduction measure. They are grown largely as a cash crop; the increasingly large town and urban population of Rwanda craves french fries.

Other features which distinguished potato from bean systems in Rwanda in the pre-war period were as follows:

(i) The varietal diversity in potatoes, both countrywide and on an individual farm basis, was not very extensive. Reports dating back 25 years show five major varieties of potato in use (Durr, 1983). National surveys conducted in 1985 suggested that a total of three to four dozen varieties were used (Haugerud, n.d.). In 1992, farmers identified 41 varieties that they had ever planted (unpublished data, PRAPACE/CIP database). Only four of these 41 varieties were sown by more than 5% of farmers.[5] Equally, mixtures of potato varieties – in this case, mixtures of improved varieties – never seemed to be very important. While work in 1985 cites the average number of potato varieties grown by any single household as 4-5 (Haugerud n.d.), research in the early 1990s suggested that the number had already narrowed to 1-3 (unpublished data, PRAPACE/CIP database).

(ii) Potato cultivation in Rwanda, unlike the cultivation of beans or any other major crop, had a strong tradition of accompanying *purchased* input use. As of 1992, 93% of farmers regularly used fungicide on their potato crop, with 69% of farmers in major potato producing areas using fertiliser (*ibid.*).

(iii) The formal sector was an important source of potato seed. Evidence suggests that by 1992 over 25% of potato farmers used improved seed from development projects or government extension, with this figure rising to 32% in areas of intensive potato production (P. Ewell, pers. comm.). A 1992 survey showed that only 4% of farmers received potato seed free, either from family or neighbours (unpublished data, PRAPACE/CIP database).

In many respects then, pre-war potato systems in Rwanda were quite unlike pre-war bean systems. Potato production was geographically concentrated and tied to purchased inputs. The formal sector played an important role in seed supply and varietal diversity was not large. However, similarities did exist. In both bean and potato systems, the 'social exchange of seed' (giving seed free) was nearly

5. All these surveys were pre-war. While the conflict did start in certain zones in Rwanda in late 1990, peace reigned in areas sampled in the survey.

non-existent and, of course, both crops were grown by small farmers as Rwanda is almost exclusively a country of small farmers.

4. KEY PATTERNS OF THE WAR

The specific effects of the Rwandan war on bean and potato production are elaborated in section 5. This section presents several overview observations about the effects of the war on Rwandan agriculture to help frame the interpretation of the findings.

4.1 Genocide and war

Rwandans themselves often distinguish between the genocide and the war *per se.* Almost all regions were affected by the genocide (although to very different degrees) but many regions, for example most of the southwest, experienced few or no direct effects of the war; fleeing populations might have passed through, but not much more.

The genocide, the worst in history, took place over about an eight to ten week period, from early April to June 1994. It broke out with the shooting down of the presidential plane on April 6, 1994. As farmers normally start planting for the second main season in January/February, most crops had been sown when the 'balloon burst'. The Rwandese Patriotic Front (RPF) gained control of much of the country by early July, just around harvest time in most areas.

The war, that is the fighting between government and RPF troops, took place at different times in different parts of the country. Indeed, whether, as of early 1997, the war is 'over' is open to interpretation. The intensive period of fighting within Rwanda ended quickly – within several weeks. Parts of the North were affected in early 1991-1992, while for much of the rest of the country, the war started in July 1994 with the southward advance of the RPF. The fleeing government troops and killing squads forced many civilians to leave their homes so as to deliver to the victor an 'empty' country. Over a million Rwandan civilians crossed national borders. Nevertheless, at least three to four times that number stayed in Rwanda.

By the time sowing took place in September/October 1994 a good measure of internal stability had been achieved although incursions at Rwanda's border still persisted. A new government structure was also firmly in place by this time. For the purposes of this paper, the first "post-war" season started with the September/October 1994 sowing.

4.2 Agricultural disruption: the broad view

While the effects of the war were very intensive for several months, those who did not flee (either from the genocide or from the RPF forces) were able to harvest much of what they had sown. Of those who did flee but who eventually returned, the majority lost one sowing season. This 'shortish duration' of the direct conflict makes the Rwandan war very different from many of the other recent wars on the African continent. For example, the Liberian war endured for at least five seasons and episodes of fighting in Guinea-Bissau lasted some 13 years (see analysis of Richards *et al.*, 1995).

The *relative* stability of the population was also reflected in patterns of residence and movement. Among those Rwandan farmers interviewed in late 1995, 90% had previously been farming in the same area and knew the micro-ecology well, including the appropriate seeds to sow. Time spent away from the homestead, which is also a direct indicator of agricultural disruption was, on average, four months, although this varied greatly by region (from 3.4 weeks in the southwest to 54.6 in the northwest). Third, and most surprising given the bombardment of media images of people on the run, was that 30% of those still farming in Rwanda at the end 1995 had not been displaced at all, not even for a single day. Evidence of this was noted in two thirds of the communes sampled country-wide by the Seeds of Hope initiative (SOH Assessment Document 8)). Family stability has important implications for varietal stability as stable families generally did not lose their seed stocks.

The degree to which farmers were, during the primary war season, able to harvest a good portion of what they had sowed is suggested in Table 2. Overall, harvest rates were significantly better that officials first estimated, with tubers generally faring better than legumes or cereals. However, there was great variability, even over short distances. The harvesting of sorghum in central Rwanda illustrates the point: some areas in the combat zone experienced total loss of their crop while farmers as little as 25 km away reaped normal harvests.

Table 2. Overall harvest rates during the primary war season, February-July 1994*

Crop	**% of farmers who harvested at least a part of what they had sown**
Beans	55
Sorghum	54
Cassava	91**
Potato	63

* A total of 883 farmers were interviewed for the bean, sorghum and cassava surveys, carried out countrywide. The potato research was concentrated in the major and minor potato zones, reaching a sample of 348 farmers.

** Cassava fields were rarely completely destroyed, but stealing was rampant (reported by 82% of cassava farmers).

4.3 Infrastructural disruption – the very broad view

It is hard to define the parameters of a 'usual' war, but the Rwandan conflict seems to have been 'unusual' in its overall aims: neither side wanted to destroy the country *per se*: their intent was rather to change the profile of the people living it. As one official reflected, post-war, 'they wanted the same envelope – but with a different contents.'

These competing goals to 'retain what was' and yet change the power (and population) balance were manifest in the patterns of infrastructural damage. Most towns looked the same before and immediately after the war, with only isolated buildings belonging to government collaborators having been blown up by victorious RPF forces. Damage in the countryside was similarly selective. Over a third of farmers living in Rwanda at the end of 1995 had no damage at all to their household or property. However, houses of the murdered and/or refugees (i.e. those not interviewed) were heavily pillaged (doors, windows, roofs ripped off) or elsewhere disassembled altogether, brick by brick. As of late 1995, 7% of farmers in Rwanda reported total household destruction (SOH Assessment Document 8). It is difficult to judge how different some of the wartime behaviour was from the norm. For instance, during countrywide interviews, a good number of farmers remarked: 'Oh the stealing was terrible,' but then continued 'but this was nothing new'.

Fields also stayed in remarkably good condition during the war period (there was pillaging, but burning was very restricted), with two general exceptions. Several agricultural areas (e.g. Rukara, in Kibungo) were extensively trampled when the huge herds kept outside Rwandan borders made their way home for the first time in thirty years. Many had fled out of the country during ethnic purges of the late 1950's and early 1960's). Second, areas of the northeast, some of which were vacated for a full two years, experienced massive vegetative overgrowth. These 'force-fallowed' sites were subsequently difficult to re-establish into working farms.

So, in sum, for such an awful war, land and infrastructural damage in Rwanda was relatively limited. Selected houses, rather than a way of life, were destroyed.

5. BEAN AND POTATO GENETIC DIVERSITY AND PRODUCTION: POST-WAR

This section draws from a series of nation-wide surveys and intensive interviews conducted during the first three post-war growing seasons. This included two seasons where there was significant emergency relief (95A and 95B), and a period when farmers were left mainly to their own resources (96A) [6] .The 1996A bean research encompassed a random, countrywide sample of 883 informants, making it comparable in scope to the Ministry of Agriculture's Department of Agriculture statistics pre-war. The 1996A potato research focused on the major and important minor production zones and reached a sample of 348 farmers. There were only two stipulations in choice of respondents: the families had to be farming and had to have previously grown the crop in question. By contrast, research during the emergency period (1995 A and B) focused on aid recipients (c. 80% of the total population).

5.1 Beans

Rate of sowing and area

In the first two major bean growing seasons post-war (1995a and 1996a), more than 95% of those interviewed sowed beans. This figure is higher than normal and reflects farmers' rehabilitation strategy of focusing on short cycle and, if possible, high value crops. Particularly during the first season, many farmers feared thinking too far into the future. For instance, relatively few invested in cassava, a crop that takes at least eighteen months to mature.

By farmer's own assessments, the area under bean cultivation post-war was fairly comparable to that sown pre-war. Forty-four per cent of farmers indicated that the surface area they were sowing was the same, while 14% indicated that they were sowing a larger area of beans because they wanted to

---

6. During the emergency period, SOH worked closely with the NGOs Care, Médicins sanș Frontières, World Vision, Swiss Disaster Relief and Catholic Relief Services. The assessments covered the September 1994 to January 1995 (95A) and February to July 1995 (95B) seasons. They monitored the impact of aid in beans, sorghum and maize.

   In the first post-war season, September 1995 to January 1996 (96A), SOH collaborated with the national programmeme, ISAR, to undertake the first intensive post-war agricultural surveys. In-depth interviews were conducted on potatoes, beans, sorghum and cassava, in all prefectures and two-thirds of the Rwandan communes (90 of the 144 total). About 1,200 households were covered. The sample size was large but also geographically dispersed in order that the micro-variations in effects of the war across small spatial distances could be captured.

rebuild stocks. Some also had access to larger areas of land (through the death of relatives and neighbours). The main reason given for diminution of bean plots (among the two-fifths of farmers who were sowing a smaller area) was lack of seed, although some also cited a greatly reduced labour force.

Seed stocks

Farmers were able to draw on their own seed stocks to a surprisingly large extent during both the 1995A and 1996A seasons; slightly under half the total quantity of seed sown came from farmers' own former harvests (Table 3). Given that many farmers had been dislocated just at harvest time, this use of 'own stock' is a welcome finding. Saved seed most often comprises well-adapted, location-specific bush bean mixtures, but may also include single varieties (local or improved) which the farmer finds productive or interesting because of certain quality attributes. Use of one's own stock encourages production stability and the preservation of local varietal diversity.

Table 3. Sources of bean seed sown (% seed sown) in Rwanda, nationwide data, seasons 1995A and 96A

Source	% Total seed 1995A (N=143 farmers)	% Total seed 1996A (N=883 farmers)
Own stock	45	40
Relief aid	28	6
Market	26	52
Friends/neighbours	<1	1
Kin	<1	<1
TOTAL	100	100

Seed donated by aid agencies (sometimes mixtures from neighbouring Burundi and Uganda and sometimes single Ugandan varieties) was of far greater importance during the 1995A season than the 1996A season. There were two reasons for this: (i) the 1995A sample was specifically targeted towards households which had received NGO relief; and (ii) the second survey was conducted during a period of agricultural rehabilitation; many who had initially fled were finding their way back home.

As had been the case pre-war, Rwandan farmers relied heavily on markets – usually the very small local ones located within a few kilometres of their homes – for seed during both the emergency relief and rehabilitation seasons. Seed gifts or borrowing of seed from neighbours and kin was negligible both pre- and post-war (Sperling and Loevinsohn, 1993).

Assessments of varietal loss and varietal reaccession

When assessing seed stocks for possible varietal erosion, it is crucial to consider the number of farmers holding local varieties across different locations, as well as aggregate figures on kilogrammes of saved seed sown. The greater the percentage of farmers holding even small quantities of local seed, and the greater their geographical dispersion, the greater the possibility for varietal diffusion and recovery. Surveys conducted during the 1995A season showed that upwards of 62% of farmers nationwide drew on their own seed stocks to some degree with at least 40% of farmers in each prefecture accessing this source (SOH Assessment Document 1).

Does this then mean that individual farmers did not lose important varieties? Not necessarily: farmers recounted selective pillaging in many areas. They claim that even their closest neighbours may have used the wartime disruption to swipe the varieties they had always coveted.

While genetic and varietal resource assessments are often done on a broad scale, for example, by country or by eco-zone, SOH work sought to conduct highly site-specific assessments. Rwandans often spend their entire lives within the radii of several '*collines*' (hills). This may mean that they are not able to access a variety which can be found only 15 km away from their home. Another reason why SOH concentrated on highly site-specific research was that the war had affected people in a very different manner even within a small geographical area.

The SOH research used three separate methods to analyse varietal loss in beans:

(i) *Overall assessments.* Farmers were asked to compare their current bean seed mixture(s) with those they had sown directly prior to the war. The aim was to get an idea of the overall performance of their current mixture and to document specific changes (both positive and negative) which may have occurred.

(ii) *Principal components analysis.* They were then asked to take their current mixture, separate out the principal components and compare this key component profile with that used just before the war.

(iii) *Tally of loss of specific varieties.* As a third check, farmers were asked to recall any varieties abandoned during the war period and *why*. Getting at this latter 'why' is very important as Rwandan farmers regularly experiment with varieties; discarding a variety is not the same as losing a variety.[7]

The results of these three methods showed that a minimal number of varieties were actually lost. Overall, 76% of farmers said that their post-war mixture was the same or better than that used during a comparable pre-war period (SOH Assessment Document 8). Eighty-four percent had all the same key components (with another 7% having two out of the three) (*Ibid.*). Further, although some farmers had lost varieties, re-accession was shown to be widely feasible.

This discussion of re-accession proved to be as important as discussion of varietal loss. Varieties can be 'lost' (for instance, poor Rwandan farmers regularly consume all their own seed) but also routinely re-accessed. Farmers' main source for re-accessing varieties was local markets, and second, friends. Knowing where a variety might be re-accessed does not, however, necessarily mean that the farmer rushes out to obtain it. There are a number of reasons why this was the case in the immediate post-war period:

(i) In many regions, farmers viewed seed price as inordinately high. They therefore refrained from purchasing supplementary seed.

---

7. A Rwandan farmer might be exposed to 100 bean varieties in a lifetime and may test about half.

(ii) Farmers complained of having to buy entire mixtures in order to get the one or two varieties they really wanted.

(iii) Farmers' priorities during this period of reconstruction simply lay elsewhere. Rather than purchase the bean varieties they had lost, farmers wanted to get a door on the house, fix the windows, and maybe pay for labour they had not needed before.

While slightly more than a third of survey respondents had lost a bean variety during the war (not necessarily a key one), most could readily re-access this lost variety (see Table 4). Only 13% of the total sample of farmers presently 'did not know' where to re-access specific varieties. This number is quite small considering that on average each farmer manages 12 varieties and that here farmers are speaking about not being able to re-access a single entry (SOH Assessment Document 8).

It is important to note that a good portion of the 'lost' varieties which farmers did *not* know how to access were improved ones, particularly the highly desired and relatively new climbing bean varieties. These varieties could generally be re-obtained only if they had already entered the local channels (markets etc.), a process that takes about three to five years. Formal seed supply channels (the central seed service and its development project outreach) were much harder hit by the war than local sources of supply. Ironically, then, while the international community was looking to restock local germplasm (given the relatively new rhetoric of biodiversity), farmers themselves felt most help was needed in getting the newer, 'improved' introductions back.

Table 4. Key bean varietal loss during the Rwandan war

Have you lost bean varieties due to the war (N=805)	% Farmers	If so, where can they be re-accessed?		
No	67			
Yes	33		% Cases (N=454)	% Farmers (N=805)
		Market	37	13
		Friends (direct)	7	2
		Friends (indirect)*	22	10
		Elsewhere	2	<1
		Don't know	32	13

* Through cross-referencing interviews, the variety has been found with a neighbour 100-200m distant. Whether the farmer can re-obtain seed has yet to be tested.

Changing varietal profiles

While there is no evidence for linking war to varietal erosion, both quantitative and qualitative insights suggest that important varietal changes have occurred over the last decade in Rwanda. This is due to two related trends: the adoption of climbing varieties in an effort to intensify production and a partial shift in varieties in response to an increased incidence of root rot.

(i) Climbing beans: Research shows a swift acceleration in the adoption of improved climbing bean varieties in Rwanda. Pre-1980, about 10% of farmers, mainly in the northwest of the country, grew indigenous varieties of climbing beans. A 1992 survey showed just over 40% of farmers spread throughout the country growing improved climbing bean varieties (Sperling and Munyaneza, 1995). The post-war survey indicated that climbing bean use was still rising. These varieties were used by 48% of bean farmers and accounted for a third of the total bean seed sown (SOH Assessment Document 8). Across the country there are zones where improved climbers are pushing out both local bush and local climbing types. For example, in the commune of Nyanyumba (Gisenyi), farmers describe their 'local' mixture as composed of three improved climbing varieties. The effects of the war in this last zone were minimal and farmers specifically noted that the local climbing mixes were dropped around 1992, as 'they were no longer productive'. The popularity of climbing beans is due to the fact that they can give two to four times the yield of bush varieties. In addition, many of the improved varieties show exceptional tolerance to root rot.

(ii) Root rot: Farmer evaluations also showed that local varieties had been dropped due to heavy disease pressure across the country, particularly in the south-centre and parts of the northwest. In some of these areas, there has been a simultaneous shift to climbers. However, in communes of the southeast farmers are still using bush beans, but experimenting with changing the varietal profile in response to disease, notably root rot. Again, this dropping of varieties has occurred in areas in which the effects of the war were minimal. Research in south-centre Rwanda (Runinya, Butare) for three years directly before the war vividly illustrates how swift the effects of root rot can be on varietal use (Buruchara n.d.). When conducting root rot trials in 1991A, the CIAT pathologist asked farmers to use their local mixtures as the control. Each farmer used her own mixture, meaning that many mixtures were used across trial sites. Farmers tended to have 12-15 varietal components in their control mixtures. In 1992A, 1992B and 1993A comparable trials were repeated but with a smaller number of farmers. In 1992, the number of varieties in use in the control fell to 3-6. In 1993A, farmers asked the researcher not to make them use their own mixtures as controls as yields from these mixtures had fallen so dramatically. Farmers preferred as local controls some of the root rot-resistant entries which the formal research system had already tested (R. Buruchara, personal communication).

It is important to emphasise that both these trends are the result of farmers' conscious strategies to combat stress. They result neither from war, nor from commercial pressures to modernise. In some cases, farmers are swapping part of their local stocks for newer varieties. In others, local varieties are being discarded in favour of other local varieties.

Absolute versus relative lack of seed

The most common stress cited by farmers in the post-war surveys was lack of seed at the household level or, more specifically, lack of money to buy necessary seed. Seed was available in the market but farmers did not have the means to *access* it and/or were angry they had to pay so much for it. This relative lack of seed in farmers' own stocks, as opposed to lack of varieties *per se* was very widespread; the problem was cited by a good portion of farmers in over half the communes sampled (SOH Assessment Document 8). Relative lack of seed is indicative of poverty rather than varietal erosion. Compensating interventions should not therefore be germplasm-based. Rather, attention should be devoted to innovative poverty-focused projects and, perhaps, selective distribution of vouchers to buy local seed.

By contrast, absolute lack of bean seed and absolute lack of varieties were rare in post-war Rwanda. These characteristics were restricted to those zones which had experienced massive killing, large transient populations, and which were ecologically marginal even before the war.

Concluding reflections: bean seed and bean diversity

Despite the much publicised civil war and genocide, the Rwandan bean varietal situation looks relatively promising. The limited varietal damage inflicted by the war is partly due to the pattern of the war itself and partly to the impressive response of the aid community which helped farmers to maintain adapted stocks. Further, one of Rwanda's main advantages, in terms of safeguarding diversity, lies in the characteristics of its existing pre-war seed channel structure. For self-pollinated crops, purchased seed from local markets was always of major importance to farmers. These markets were quickly re-established post-war as seeds of beans and sorghum, for example, could be restocked locally. Even during the first season post-war, surveys showed markets supplying as much seed as relief aid agencies. Restocking through neighbours and friends was never particularly important in Rwanda (at least not for the current generation). This meant that ruptured social relationships had little effect on bean seed systems.

Varietal diversity in beans remains impressive in Rwanda. SOH workers during the first post-war harvest of January 1995, were able collect some 1,300 different phenotypes from a very restricted sample of about 150 households and 20 markets (S. Beebe, personal communication). The situation is, however, far from static. Farmers' strategies to intensify production and combat root rot are resulting in unusually dynamic varietal profiles for beans at the current time.

5.2 Potatoes

Rate of sowing and area

The case of potatoes is very different from that of beans. Only 84% of active potato farmers interviewed had sown any potato tubers during the October-November 1995 (1995B) period. When comparing the surface area they had sown to potatoes during 1995 with that sown during a comparable pre-war period, a full two-thirds of farmers said they were planting a smaller area of potatoes post-war.[8] Ten percent of farmers were not planting at all. When asked about the reasons for the contraction of potato areas, two-thirds responded: *lack of seed.* Both absolute lack of seed (none available either at home or in the market) and relative lack of seed were cited.

Seed stocks

Virtually all of the potato seed sown in the 1995B period came from farmers' own stocks (55%) or the market (42%). Exchange of potato seed among neighbours or friends had always been rare. Unlike beans, however, use of what farmers termed 'local varieties' also proved to be negligible. Only 3% of farmers claimed to have sown local varieties even pre-war, and during the first post-war season, only 5% of the seed sown was local.

Aid agencies were not a source of potato seed. Tubers are hard to move on a large scale, and potato is not a priority crop nationwide, hence the agencies focused their efforts on other crops. They did,

8. Pre-war farmers often sowed about 100-200 kg of potatoes, an investment many could no longer afford

however, contribute substantial technical assistance to help reestablish domestic potato seed multiplication facilities (both through selected farmer-multipliers and the parastatals).

Varietal loss, seed loss and reaccession possibilities

With potatoes, the issue of varietal loss never loomed large. Three officially-released varieties, Cruza, Sangema and Mabondo (the first two of Mexican origin, the latter Rwandan), dominate potato fields, accounting for 86% of the total seed sown (SOH Assessment Document 9). Indeed, in the south-central region (Gikongoro) Cruza alone accounted for 86% of all seed sown.

Moreover, most farmers interviewed had never even experimented with a significant range of varieties. Table 5 shows the number of varieties sampled farmers currently use and the number they have *ever* tried. Most of the sample had tested both Cruza and Sangema, with Mabondo, Montsama and Muhabura having had lesser exposure. All are improved cultivars. Few other varieties have ever reached more than a handful of farmers. This range is limited considering both farmers' high interest in potato production and their propensity for lively varietal experimentation in crops such as beans.

Obtaining adequate volumes of desired varieties of potato seed, rather than accessing the germplasm itself, was a real problem for many farmers in the post-war emergency and rehabilitation seasons. Overall, two-fifths of farmers said they had lost valued seed of certain varieties, with the region of Byumba, where farmers were absent for about two years from the potato growing zones, being particularly badly hit. Fifty-six percent of farmers there experienced significant seed loss.

Part of the problem in potato re-acquisition has been the relative dependence of farmers on formal, external sources of seed. In the last ten years, farmers have come to rely heavily on development projects and the national agricultural research system (NARS) for sourcing clean seed and for accessing the few new varieties, often at subsidised prices. The war disrupted this supply as early as 1991-92 because the NARS parastatal responsible for bulking up was located in one of the war's first fronts. Many development projects began to phase out their activities during 1993.

If restocking possibilities are considered on a case by case basis (that is, where can 'x' variety be found?), the scenario of Table 6 looks relatively grim for the potato zones as a whole. In almost half the cases, it is not clear where farmers are going to find seed of potato varieties – although about a quarter of the varieties 'lost' can be re-accessed on the market.

However, when the data is analysed from the farmer perspective within each region, it becomes clear that markets in all regions are good sources of material, and that relatively few farmers cannot find material. The 'I don't know where to find x' figures are inflated because the same farmer is indicating that she/he cannot find two or three different varieties. Again, the exception is Byumba, where almost a third of the farmers are not sure where to re-access potato varietal material. There may be an absolute lack of seed in Byumba. In other regions, however, the problem of obtaining potato seed was due to lack of money. This can therefore be considered as relative rather than absolute lack.

Post-war surveys showed that most Rwandan potato farmers (56%) had used fertiliser at some time in their potato farming careers.[9] However, in some regions of the country, such as south-centre,

9. The term 'fertiliser' generally refers to different combinations of NPK.

farmers complained that supplies of fungicide and fertiliser dried up in 1992/93. Potato production was tied to use of these purchased inputs and one reason why farmers did not look for potato seed was that the accompanying inputs they deemed to be essential were no longer available.

Table 5. Number of potato varieties ever tried and currently used by Rwandan farmers: 1996

Number of varieties ever tried	% of potato farmers (N=347)	Number of potato varieties currently used	% of potato farmers (N=293)
1	17	1	51
2	31	2	34
3	26	3	10
4	16	4	4
5	5	5	<1
6	2	6	<1
7	1	-	-
8	<1	-	-
10	<1	-	-

Table 6. Sources where farmers can find lost potato varieties, across potato zones (N=130 farmers with 206 cases of 'loss'): 1996

Source	% of cases of 'loss'
Market	27
Friend	3
Development Projects	12
Other	5
NARS	5
Don't know	47

Input use: fertiliser and fungicide

Twenty-three per cent of those who had previously used fertiliser were no longer doing so in the immediate post-war period because of lack of availability (the majority of cases) or high cost. The same patterns can be observed for fungicide, use of which had been even more prevalent than that of fertiliser, two-thirds of farmers having applied fungicide at some point, but only one third of these still applying it (see Table 7).

Table 7. Evolution in use of fungicide and fertiliser by Rwandan potato farmers*

Use of fungicide: last time (N=273)	% of farmers	Use of fertiliser: last time (N=194)	% of farmers
1982	<1	1981	<1
		1983	1
		1988	1
		1989	2
1990	8	1990	11
1991	5	1991	6
1992	5	1992	4
1993	11	1993	12
1994	38	1994	40
Continued use in 1995–96	33	Continued use in 1995–96	23

* Sample focuses only on those who have tried fungicide or fertiliser.

Concluding reflections: potato seed and potato diversity

Potato production in Rwanda was hit particularly hard by the war. Overall production is significantly down with two-thirds of potato farmers sowing a smaller area than at a comparable period before the war. Availability of seed is a general problem (although the northwest zone is slightly better off than the others). Fungicide and fertiliser use have also fallen dramatically with only 33 % and 23 % respectively of those who have ever tried these inputs still using them.

As was the case for beans, the war in Rwanda did not result in a problem of potato varietal loss *per se*. However, whereas the main problem in bean systems was a relative lack of planting material (farmers could not pay for seed at market prices), potato farmers also faced an absolute lack of seed (as in Byumba) or an absolute of clean seed. Pre-war, production of clean potato seed had been quite highly centralised. It had been produced either within the national research programme in the northwest or in certain development projects supported by expatriate funding. Neither source proved sustainable during the wartime and immediate post-war period.

Potato farmers appear then to feel a lack of control over the process of rejuvenation. They have done little to rebuild potato production when compared with their active strategies for rebuilding bean production.

6. LESSONS ON SEED SECURITY AND VARIETAL SECURITY ASSESSMENTS

In terms of thinking about seed security, varietal security and war, the Rwandan contrast between bean and potato systems offers a series of insights about the effects of acute stress on crop production. They are summarised as follows.

6.1 Post war seed system and varietal diagnosis

When aiming to assess the effects of war on varietal diversity, researchers should focus equal attention on the *seed channels* which can re-supply germplasm and on the germplasm itself. The issue is not whether a farmer is using a particular variety (or set of varieties) at a single point in time, but rather whether she/he can re-access it if desired. Varietal loss is a dynamic process, but so is varietal accession. **A key for understanding the processes of diversity in times of acute stress lies in the analysis of seed channels.**

A war does not affect the functioning of all seed channels in the same ways. There may be variations according to type of system (e.g. formal vs. farmer) and according to the type of crop within any particular channel. Unexpectedly, in Rwanda, the formal seed channels took the hardest beating in the war years, while farmers' own seed systems (for beans and sorghum, at least) proved to be surprisingly resilient. For this reason, analysis of one crop cannot necessarily to extrapolated to another. **Those interested in varietal diversity and rebuilding agriculture after calamity need to be prepared to adopt an in-depth, multi-crop focus and need to give significant attention to both farmer and formal seed channels.**

It is imperative to try to distinguish between farmers' absolute versus relative lack of seed and, similarly, absolute and relative lack of varieties. Absolute lack implies a true scarcity of varieties or seed in a region. Remedial action in such circumstances should focus on re-introduction or interventions to build up seed production capacity (as was the case with potatoes in Rwanda). Relative lack of varieties/seed implies problems with *accessing* seed, not absence of seed *per se*. Seed (and varieties) may be generally available but not within farmers' specific means. In such circumstances, interventions to counter 'seed deficits' or 'varietal erosion' should not be germplasm-based. Rather, innovative poverty-focused projects (to generate income) and, perhaps, selective distribution of 'seed vouchers' (to buy local seed) should be considered.

6.2 Post-war germplasm re-introductions

The Rwandan analysis clearly shows the dynamism in varietal use. Varieties in use today may differ markedly from varieties deposited in genebanks only a decade ago, even in the absence of turbulent events or strong promotional policies to use improved material. Opening up the genebanks to restore on-farm varietal diversity will not, therefore, automatically deliver an adapted or farmer-acceptable product in times of stress.

The present-day trend for promoting biodiversity has led many concerned activists, environmentalists, and farming systems specialists (among others) to advocate as a near panacea the restoration of farmer germplasm to its original sites of use. Careful documentation of bean varietal use in Rwanda shows that such an approach may not always be to farmers' benefit, even in low input situations with resource poor farmers. The rapid emergence of root rot in Rwanda and the change in varietal use in response to this provides just one example of the relative lack of dynamism in genebanks as opposed to farmers' fields.

If germplasm is to be re-introduced, it should, as far as possible, resemble that which farmers were using directly prior to the emergency situation (assuming that the agro-ecological context was a stable, viable one). This implies that a frontline preparedness strategy for emergency situations should focus in the first instance on understanding which farmers are sowing which varietal material and why (see also ODI, 1996).

It is fortunate – for both Rwandan farmers and the world community – that Rwandan germplasm seems to be in relatively good shape. However, SOH did, during its post-war research, prepare a contingency plan for reintroduction of bean germplasm, as the challenges of site-specific restoration became more and more apparent.

The components of this strategy may be of interest to other potential interveners:

Step 1: Landraces and improved varieties were multiplied in important initial quantities.
Step 2: The germplasm was to be colour-coded by general attitudinal adaptation.
Step 3: Testing of germplasm was to be decentralised to farming communities themselves, partly with the support of NGOs. Farmers themselves were to be given a diverse range of germplasm to screen on-site.

In the end, the plan never needed to be executed. The loss of germplasm itself was not as great as had been anticipated because of the particular patterns of the Rwandan war. Further, local markets were able to re-supply reasonable quantities of seed of locally adapted varieties when farmers did need to re-access specific varieties.

The plan, however, did take as a central tenet that **'germplasm is only actually 'restored' when it starts to grow again and to evolve *in farmers' fields***. This means that germplasm restoration means more than transferring material from one national (or international) genebank to another. As even in normal circumstances, many of the poorest are never reached by formal seed systems, in stress situations, the challenge will magnified to aid the most needy. Innovative strategies will be needed to get locally adapted germplasm to specific communities. Participatory plant breeding and variety selection (Step 3; above) may provide several viable options.

7. SELECT LESSONS FOR DELIVERING SEED AID AND GIVING SEED SYSTEM SUPPORT

The SOH work in Rwanda also implicitly identified a number of 'better practice' strategies for delivering seed aid or support to seed systems more generally. Some of the insights are summarized below for the phases of disaster mitigation and disaster preparedness.

7.1 Disaster Mitigation

Massive amounts of food and seed aid (particularly of beans and maize) arrived in Rwanda within two to three months of the ending of the genocide. The scale of the relief effort was impressive: at one point there were thirty NGOs distributing seed alone. (This in a country about the size of Switzerland). **The swift food and seed aid response of the international community was probably the single most important factor in allowing farmers to save their own adapted seed stocks.** The quick restoration of relative security for those who remained was also key: farmers sensed that what they would sow in September 1994 would indeed be harvested--so they were willing to continue to pursue agricultural activities.

Delivery of variety-sensitive seed aid proved extremely cost-effective. The definition of variety-sensitive aid encompasses the notion of both adapted and farmer-acceptable varieties. The range of varieties on offer should also be diverse enough to meet farmers' critical needs (e.g. varieties for different foods; different planting periods).

In the Rwanda case, two aspects of this variety-sensitive seed aid deserve particular mention:

- For beans and sorghum, crops characterized pre-war by broad varietal diversity and sowing in varietal mixtures, aid workers adopted a conscious strategy of distributing mixtures. On-farm evaluations showed that distributing mixtures (and a range of varieties) a) promoted production stability; and b) permitted farmers to recover or select out "best varietal bets" (e.g. farmers were able to select out varieties which later performed well).

- Several aid agencies spent considerable effort targeting adapted materials to specific zones. Well-informed suppliers bought seed from comparable agro-ecological regions in adjacent countries. For beans, suppliers looked to central and southern Uganda and to Burundi. For sorghum, much of the supply from moved from internal Rwanda stores or from southern Uganda. Médecins sans Frontières even tried to target mixtures within Rwanda and they distributed four different varietal blends just for the southern area: three from Uganda and one from Burundi. Yield data clearly showed the effectiveness of targeting.

Both these aspects of disaster mitigation suggest that there should be a strong role for germplasm specialists in shaping relief aid.

7.2 Disaster preparedness

Farmers use both informal (a.k.a. farmer, local) seed channels and formal ones. As disruption may affect these channels differently, both need to be the explicit focus of a preparedness strategy. It is important to understand how both sets of seed systems (informal and formal) operate in normal times. Only then can one properly diagnose if, and in what way systems are functioning (or malfuntioning) in a stress period. Seed systems have both technical dimensions (e.g. type of variety, phytosanitary aspects) and social ones (e.g. those based on exchange, embedded with ritual significance). A holistic diagnosis, central to a preparedness strategy, has to ask such questions as: who uses which specific channels? (e.g. purchase from local markets, gifts of seed from neighbors) for which crops, varieties and quantities of seed? when can each channel be used? with which restrictions? Understanding how the system normally operates provides the base for pinpointing what kind of seed system support may be needed in a stress period. It also provides important insights into how not to undermine what may be functioning aspects of seed systems in emergency periods.

A prerequisite for delivering variety-sensitive seed aid is having specific knowledge of which varieties can be used where. Databases which can match varieties to environments proved useful in the Rwandan case for all crops delivered. This involved two basic types of 'matching': (i) agro-ecological matching of varieties to physical environments, and; (ii) cultural and economic matching of varieties to known farmer preferences.

Regulations need to be sufficiently flexible to move farmer-acceptable seed quickly. This may involve movement of seed across several international borders; movement of varieties which are not officially released, and in some cases, movement of seed which is not fully certified. All three aspects proved crucial to delivering useful seed during the Rwandan crisis.

8. REFERENCES

Buruchara, R. (n.d.) 'Development of management options for bean root rots: lessons learnt from research in the Great Lakes Region.' Paper presented in the *Huitième Séminaire Régional sur l'Amélioration de haricot dans la Région des Grands Lacs*, 5-8 November, 1995. Kampala, Uganda: International Centre for Tropical Agriculture.

Cishahayo, D., Loevinsohn, M., Sperling, L. and Mukagatera, V. (n.d.) 'Diversité génétique et dynamique des mélanges de haricot.' Karama, Rwanda: Institut des Sciences Agronomiques du Rwanda, manuscript.

Durr, G. (1983) 'Potato production and utilization in Rwanda.' Social Science Department, *Working Paper 1983-1*. Lima: International Potato Centre.

Haugerud, A. (n.d.) 'Farmers' criteria in selection and use of potato cultivars.' Paper prepared for Regional Potato Germplasm Course, June 10-25, 1985. Nairobi: CIP Regional Office.

Iyameremye, A. (n.d.) 'Le secteur agricole durement touché.' *Traits d'union, Rwanda 5.*

Lamb, E.M. and Hardman, L.L. (1985) *Final report of: Survey of bean varieties grown in Rwanda. January 1984-June 1985.* AID-Rwanda Local Crop Storage Cooperative Research.

MINAGRI/UNREO/PNUD/FAO. 1994. Compte-rendu de réunion, 16 décembre 1994: Opérations d'urgence et réhabilitation agricoles.

Overseas Development Institute (ODI) Seeds and Biodiversity Programme. (1996) 'Seed provision during and after emergencies.' Relief and Rehabilitation Network *Good Practice Review 4.* London: Overseas Development Institute and Euronaid.

Scheidegger, U.C. (1993) *The diversity of bean varieties in Rwanda. Bean Programme Annual Report 1993.* Cali: Colombia: International Centre for Tropical Agriculture.

SOH Assessment Documents:

1 Assessment of bean seed use 1995A: Summary Report
2 The impact of war on plant genetic resources: bean varietal assessment in Rwanda 1995A: Summary Report
3 Assessment of bean and maize seed use 1995A in CARE zone of action
4 Assessment of bean seed use 1995a in World Vision zone of action
5 Assessment of maize seed use 1995A in World Vision zone of action
6 Assessment of maize seed use 1995A: Summary report
7 Assessment of sorghum seed use 1995B in Care zone of action
8 Rwanda nationwide household survey on the status of the bean, sorghum and cassava crops 1995: The impact of war on production and varietal erosion.
9 Rwanda household survey on the status of the potato crop, 1995
10 Executive summary of Seeds of Hope (SOH) socio-economic analyses in Rwanda: the impact of the war on agricultural production

(All are deposited at: Ottawa: International Development Research Centre)

Sperling, L. and Loevinsohn, M. (1993) 'The dynamics of improved bean varieties among small farmers in Rwanda.' *Agricultural Systems* 41:441-453.

Sperling, L. and Muyaneza, S. (1995) 'Intensifying production among smallholder farmers: the impact of improved climbing beans in Rwanda.' *African Crop Science Journal* Vol. 3(1).

Voss, J. (1992) 'Conserving and increasing on-farm genetic diversity:farmer management of varietal bean mixtures in Central Africa.' In: J.L. Moock and R.E. Rhoades (eds). *Diversity, farmer knowledge and sustainability*. Ithaca and London: Cornell University Press.

Background Papers

CHARACTERIZING DISASTERS

François Grunewald, Agronomist
Groupe Urgence-Rehabilitation-Developpement, France

1. INTRODUCTION: UNDERSTANDING THE DIVERSITY

One of the sad features of today's world is that no place on Earth can be said to be 100% free of disaster risk. Disasters have affected the African, Asian, European (Central, and Eastern), south American continents. Most of the agro-ecological zones have been touched, from the arid sub-saharan Northern Mali and Horn of Africa to the humid central Africa and the monsoon belt is Southeast Asia. From the sea side of Sri Lanka and East Timor to the mountainous Hazarat in the Hindhu Kush ranges, from the immense low plains of Congo to the Nil-Zaïre highlands in Rwanda. Countries with various levels and types of development and social organisation have suffered: from manual subsistence agriculture to highly sophisticated mechanised agricultural systems, socialist and market oriented economies alike.

Disaster situations are divers. Disasters of natural origin or cause have reached unprecedented scales, since more and more marginal and risk-prone areas are inhabited and put under cultivation by a population growing by the day. Fragile ecosystems and agrarian systems are under enormous pressures from the globalisation of the world economy.

Internal conflicts and international wars are also causing immense distress for rural populations. Devastated villages, burned houses, destroyed, non harvested or even unplanted crops are common occurances. Displacements of various scales and for variable duration are also some of the results and aggravating factors of these man-made disasters. This is the century-old story of farmers, famines and war. As said by the well-known agronomist René Dumont "war is often peasants' assassination!". But the end of the Cold war, instead of delivering peace to all, brought about new waves of wars. Internal conflicts, specifically targeting civilian populations, are more than ever on the rise.

Agriculture is for many of the Planet inhabitants at the centre of the daily preoccupations. The recent disastrous floods in China and Bangladesh, the ecological and economical crisis in Indonesia, the still ragging war in post Dayton former Yugoslavia (Kosovo), the never ending conflicts in Angola and Afghanistan, all these events have important consequences on agriculture and life in the rural world. When agriculture is disturbed, food becomes scarce in the rural and urban markets, prices rise and the whole economy is put under jeopardy. Access to means of production, and more specifically to seeds and other planting materials, is often at stake.

But the degree of complexity is even higher. The "simple disaster" does not exist anymore (if it has ever done so). The largest humanitarian crises of the recent years have all been a combination of factors : war and drought (Mozambique in 1992), war and flood (Somalia in 1997-98), war and locust infestation (Ethiopia in 1986-87), massive displacements of populations for dam building and flood, etc.

A very specific complex feature is when the disaster strikes in a situation of transition: transition from war to peace (Mozambique, Angola, Bosnia-Herzegovina); transition from dictatorship to democracy (Haïti); transition from socialist central planning to market economy (North and South Caucasus),etc. Then the possible state-related decision makers are often in a weak position to take action, if even they exist. They are often sometimes inexperienced in many subjects and under a lot of pressures, many of them mainly of a pure political nature. Agriculture, restoration of the productive capacity of the farmers, conservation of the national phytogenetic resources are normally not very high in the agenda.

Civil society (local NGO, professional associations, area specific organisation, etc.) has often been partly decimated during the crisis or is hiding underground awaiting for all the carts to be redistributed. These two key players in times of crisis are, therefore, not available, thus jeopardising the chance for a proper identification of the real needs and an appropriate appraisal of the local capacities. This affects all humanitarian intervention, and more specifically the agriculture recovery programmes.

Failure in restoring productive farming systems during conflicts and/or post-conflict or post disaster periods have usually had dramatic negative impacts on human beings and social systems. They have even often affected the chance for a durable solution or/and a protracted peace. Therefore, it is of paramount importance to improve means and methods to properly respond to the agricultural needs of these troubled times in highly complex and heterogeneous contexts.

The approach on which this paper focuses on the recognition that what counts at the end of the day is the understanding of the complexity and of the diversity. Therefore, the reader should not expect the description of "an average disaster situation" with the establishment of "an average best practice". There will be no simple set of synthetic indicators. Thus, the strategy chosen to characterise disaster situations and identify the most useful variables is to go through the establishment of fine-tuned typologies :

- Typologies of the disasters (section 2);
- Typologies of the affected agricultural contexts (section 3);

Then, these two typologies will be crossed in order see how the two types of factors inter-act (section 4). This exercise, relevant and operational both in terms of analysis/understanding and in terms of prediction, planning and action.

It is only when this identification of the factors of variation has been done, when the interactions are analysed and correlations identified, that appropriate responses can be designed and implemented.

2. DISASTERS: FACTORS OF DIFFERENTIATION

Disasters are very diversified and take many different forms. Equally different are the way these disasters affect rural communities and more specifically their "seed system[10]". Among the key variables, five seem to play a very special role:

- <u>The nature of the disaster.</u>

First of all, the nature of the disaster will play a very important role. Indeed, natural disasters and man-made disasters may not affect, in the same manner and with the same magnitude, a given context. With a more sophisticated approach, we can also say that the objectives of a war themselves will lead to different types of impacts . International wars aimed at territorial or political gains will not have the same overall repercussions nor will they impact existing farming systems the same way as internal ethnic war where the objective of the fight is to eliminate the "other group", the "different

[10] The « seed system » comprises all the activities related to seeds : from production to trade through selection, local or industrial processing and storing. The way disasters affect agriculture in general and this seed system in particular will be treated in sections 3 and 4.

people". In addition, disasters, be they natural or man-made, will have very different consequence whether or not they are followed by large scale and protracted displacements. It is, therefore, of a paramount importance to identify the nature of the disaster, both by its causality and by its external manifestations. The grid presented on the following page (Table 1) lists the main types of disasters, their main characteristics and a set of indicators that can be used to characterise them both from a "causality" (predictive value) and a descriptive point of view.

Unfortunately, as it will be shown below, "simple cause disasters" are very rare. Many factors often interact to transform a crisis that could be dealt with into an "humanitarian disaster". The tools to analyse, characterised and categorise these disasters have to be found in many different disciplines: climatology, vulcanology, geology, political sciences, economy, ethnology, anthropology, biology, ecology, etc. Often, multidisciplinary approaches are needed, due to the complexity of the disasters and of their causes.

- The predictability of the phenomenon and the speed of disaster's onset;

Disasters are seldom totally unknown and unexpected factors. They often have occurred at least once in the period that human memory or local traditions can keep track of. Armed with this historical background, neither farmers, nor the local administration (whatever it may be) remain passive actors in front of possible or actual disasters. In risk-prone areas, farmers have often developed sophisticated early warning mechanisms based on the observation of indicators perceived as having an high predictive value. Rural societies have also elaborated complex risk minimisation strategies and disaster mitigation: use of various varieties and crops, spreading of the agricultural activities over a longer period of time, cultivation of different locations within an agro-ecosystem, establishment of multi-years food and seed stocks, etc., are among these strategies.

In the face of climatic events , the agricultural authorities often try their best to establish contingency plans: dams can be filled up with reserves of water ready to be release at the peak of the drought in order to save fish, to make irrigation possible or to restore the soil moisture of zones normally carrying water-receding crops. When confronted with war or related events, farmers try to hide seeds or animal flocks. While most of the observors recorded the high degree of depletion of the herds of the war affected Bar-El Gazhal and Jongley regions of south Sudan, very few have noted the local transversal migrations of the flocks towards Equatoria and the even Kenyan Turkana region.

But even if these national actors are often trying to get ready one way or another when the disaster strikes, the implementation of these measures is unfortunately only possible when there is enough advance notice. When the flood comes in a matter of hours or when the attacks are sudden and unexpected, there is vary little that can be done.

- The phenomenon's duration;

Rural economies are just another system of capitalisation and accumulation of means of production. They can take the form of livestock (many pastoral and agro-pastoral societies), food stocks (Sahelian systems), investment in the labour intensive modification of the agro-ecosystems (terracing in south-east Asia, Hindu Kush and Mediterranean regions) or accumulation of more and more costly equipment. Under the tough pressure of a disaster, these rural economies can be affected by the reverse "decapitalisation" process. The longer the duration of the disaster, the more acute and abrasive is this "decapitalisation", which can at the end lead to total or near total disappearance of the farm and destitution of the farmers. Rwandan experience has shown that the degree of affectation of a system depends heavily on the duration of the direct exposure of this system to the disaster, in that case war hostilities on the one hand, and actual displacements on the other hand. In general, the longer the duration, the more serious is the destitution process. But simultaneously, the more creative

Table 1. Determining the nature of disasters : a typology for identification

Categories	Description	Types of predictive Indicators	Types of descriptive indicators	Means of verification	Remarks
Natural disasters	Drought	Climatic indicators	Climatic indicators, Nutritional indicators, Socio-economic indicators	Climatic data Satellite imagery	For most of these types of disasters, frequency analysis of climatic data and historical review of the possible repercussions of the disaster on the situation can often be pre-established. El Nino is a perfect case of a predictable event with all the signals for early warning being perfectly known.
	Floods	Climatic indicators	Climatic indicators, Nutritional indicators,	Climatic data Satellite imagery	
	Volcano eruption	Tectonic indicators Behaviours of fauna	Seismologic indicators Agro-ecological indicators Socio-economic indicators	Geotectonic monitoring Fauna observation Satellite imagery	
	Earth quake	Tectonic indicators Behaviours of fauna	Seismologic indicators Agro-ecological indicators Socio-economic indicators	Geotectonic monitoring Fauna observation	
	Typhoons	Climatic indicators	Climatic indicators, Agro-ecological indicators	Climatic data Satellite imagery	
	Tsunami	Tectonic indicators Climatic indicators	Climatic indicators Socio-economic indicators	Geotectonic monitoring Climatic data Satellite imagery	
	Locust infestation	Climatic indicators Dynamics of the locust population	Agro-ecological indicators, Nutritional indicators, Socio-economic indicators	Climatic data Satellite imagery, Ecological surveys	
	Land slides	Climatic indicators Tectonic indicators	Agro-ecological indicators, Nutritional indicators, Socio-economic indicators	Climatic data Satellite imagery Geotectonic monitoring Risk mapping	
	Large scale outbreak of fire	Climatic indicators Ecological indicators Signs of fire	Agro-ecological indicators, Nutritional indicators, Socio-economic indicators	Climatic data Satellite imagery, Ecological surveys	Often linked to combination of natural (El Nino for instance) and socio-economic factors (land availability)
Man made disaster	International war	Geopolitical indicators Historical indicators Economic indicators Demographic indicators	Military indicators, Surgical indicators, Demographic indicators Socio-economic indicators, Nutritional indicators,	Geopolitical analysis Economic analysis Review of history	For most of these disasters, the roots have often to be sough for in the History of the area and of the people inhabiting them.
	Political internal war	Geopolitical indicators Historical indicators Demographic indicators	Military indicators, Surgical indicators, Demographic indicators Socio-economic indicators, Nutritional indicators,	Geopolitical analysis Economic analysis	
	Ethnic conflict	Political indicators Historical indicators Economic indicators Demographic indicators	Military indicators, Surgical indicators, Demographic indicators Socio-economic indicators, Nutritional indicators,	Geopolitical analysis Economic analysis Review of history	
	Results of refugee influx	Demographic indicators Ecological indicators	Military indicators, Surgical indicators, Demographic indicators Socio-economic indicators , Nutritional indicators, .	Geopolitical analysis Economic analysis	Often the consequence of one or two other causes (combination of various types of disasters)
	Failure of economic policies	Economic indicators Geopolitical indicators Historical indicators	Demographic indicators Socio-economic indicators, Nutritional indicators,	Geopolitical analysis Economic analysis Review of history	Here again most of the systems have been tried in hi
	Political sanctions	Political indicators Historical indicators Economic indicators Demographic indicators	Military indicators, Public health indicators, Demographic indicators Socio-economic indicators, Nutritional indicators,	Geopolitical analysis Economic analysis Review of history	
	Technological disaster	Tectonic indicators Ecological indicators Climatic indicators	Agro-ecological indicators, Nutritional indicators, Socio-economic indicators	Technological analysis, Risk analysis/mapping, Ecological surveys	Either link to out-of date (nuclear for instance) or untested technologies (for instance the OGM)

are the adaptive survival mechanisms. The timing of the point where the destitution line meets the creative survival line varies from one group to another, between and within disasters. This timeframe can make the difference between a human catastrophe and a simple crisis. However, some disasters are just decades long stories. In certain contexts, protracted sufferings and unbalance tend to become the norm rather than the exception. In these contexts, there is no transition from war to peace, but just long lasting grey periods when it is not clear whether the war is on or off, and who won. These never ending crises heavily weight on the existing economic systems in general and on farming systems in particular. The very existence of unaffected seed stocks is sometimes the result of a near miracle, while in many instances it is the result of farmers's highly elaborate survival mechanisms. The other elements related to "time" will be discussed in a section below.

➢ <u>The number of interacting factors :</u>

It is often a "chicken and egg story". Why have the Mozambicans suffered so much from the 1992 drought? Is it because two decades of civil war have affected so much the survival capacities of the population that it cannot cope anymore with a drought? Or is that because the drought brought down the food production and the agricultural cycles that the cumulative effects of the ongoing civil war have reached the maximum level the population can endure without severely encroaching its survival capacity? In specific cases, when famine is used as a weapon, then forced displacements, destruction of crops and burning of stocks of food staple and seeds can bring a population to a state of vulnerability that could turn into a disastrous situation.

Of course, demographic characteristics of an area or of a population can reinforce certain factors and transform them into a disaster-accelerating factor. In areas of Northern Mali, the fragile arid environment can sustain only a certain number of people per km². If the density doubles because of events in another region, then the agricultural resources, water, pastures and dry fodder for the livestock and other wild resources (fruits, wood, berries, wild cereals, etc.) will get depleted. This will sow the seeds of the next disaster. In highly populated areas, a limited event leading to only an increase of 10 % of the population can have enormous repercussions.

It has been mentioned that the transition between socialist planned economy to the new mechanisms of the market-driven system is another factor that can interact with arising conflicts. This has been seen in Aizerbaijan, Armenia, Georgia, Tchetchenia, Tajikistan, etc. Norms and references tend to disappear during these troubled times. If another strata of unbalance, such as a political crisis or even a conflict show up in the picture, then the probability is high to see the whole situation turning into a disaster with incalculable consequences. The rural world is soon affected, and so is the whole agricultural system and, its seed component.

➢ <u>The degree of reliance of the agrarian systems on external means of production, especially seeds.</u>

In large parts of Africa, subsistence agriculture is often based on family based production and reproduction of the planting materials (seeds, cuttings, roots, etc.). Farmers are not really depending on the production of these planting materials outside of their farms. Therefore, the supplies do not depend on the functioning of trading lines or mechanisms. On the other hand, when the disaster has totally destroyed local seed stocks, farmers are in a more complicated position to get replenish their seed stockes. In the Caucasus region, or in former Yugoslavia, farmers have been for decades depending on specialised seed production units for their supplies. If the economy does not enable the farmers to access these planting materials or if a frontline replaces what use to be a commercial interface, then these farmers are confronted to a very serious challenge: how can they, in a matter of days or weeks, find an alternative seed supply? How will they become able to start their own seed production? However, in these more sophisticated systems, there is a better likelihood that seeds may be available somewhere, or, at least, that the genetic material at the origin of the commercial varieties is still kept in one or another seed centre.

3. HOW DO DISASTERS AFFECT AGRICULTURAL ACTIVITIES AND BIODIVERSITY

These various factors and their repercussions in terms of possible strategies will be studied in the following chapters. The way farming systems are affected will be described with more details. There will be an attempt to "zoom" at three levels :

- Rural economy of the affected area;
- Functioning of the farming systems and food security systems; and
- The seed system (keeping, selling, buying procedures).

Agricultural production is an process by which photosynthesis is transform into useful means for the survival of human beings. It requires a certain number of factors of productions (land, manpower, capital) and inputs (seeds, water, tools, inputs coming from the industrial world, etc.) depending on the level reached by the productive forces. Disasters affect the whole agricultural cycle. They affect access to factors and means of production as well as the final results of the productive cycles: food, stocks of raw material for handicraft or other activities, products for the market, etc. This also affects the way farmers and non farmers use natural resources, including food and seeds stored in granaries or still in the fields. They can also put the very existence of natural resources in danger (napal or other chemical weapons in the rain forest, mines in pastoral lands, etc.). Disasters can even jeopardise the reproduction of the fertility of the ecosystems. In Rwanda, the disappearance of a large part of the livestock is causing direct damages on the structure and texture of the sloping fragile soils of the "country of the thousand hills".

A series of factors affects more crucially the agricultural processes:

- The "time factor";
- The "resilience factor";
- The "Global context factor"; and
- The "Aid factor".

3.1. The time factor

One of the first factors to study is the "time factor" : This time factor has to be subdivided into three components : time of occurrence, duration and frequency.

- Time of occurrence :

The question to answer is: when did the disaster take place compared with the agricultural, pastoral and non-farming calendars (see example in Table 2)? Indeed, the consequences will be different if it takes place before sowing, while the crops are growing, at harvest time or when the grarraries are full.

Table 2. **Somalia, the April 1992 war in Baidoa and the central sorghum belt**

Months	F	M	A	M	J	J	A	S	O	N	D	J
Seasons		Gu season							Dehr season			
Sorghum crop calendar		Sowing	Weeding			Harvesting	Storing and consuming		Sowing	Weeding		Harvesting
Disaster		War	Insecurity									
Effects and impacts		No plantation, Food and seed stocks destroyed or consumed	The little that was sown could not be weeded and taken care of.			Yield limited or non-existent stocks	Availability of seed extremely limited					

One can assess from Table 2 above that war activities would have had a very different effect if they would have taken place at another time of the crop calendar. In each situation, the exercise of establishing the calendar of the various productive and non productive activities and the positioning of the disaster vis-à-vis this calendar should be compulsory as a methodological tools in the diagnosis of the situation and of the planning of any activities. This is unfortunately seldom done. Many setbacks in humanitarian programmes have their origin in the disconnection of aid activities from the local calendar of planting, harvesting, moving the flocks from seasonal pastures to other areas, etc.

The military offensive that took place in the spring 1998 in Afghanistan, together with the earthquake affecting the same year this poor and unlucky country, have both taken place at the time when farmers where preparing for sowing while food stocks were nearing their end (beginning of the lean period). Repercussions would have been less dramatic if the same events had taken place at harvest or post harvest times when people where well nourished and food stocks at their peak.

We could add many examples to this list. Many attacks have taken place in former Yugoslavia just at harvest time. Those who managed to harvest were often not the ones who had planted. Thus, the socio-economic conditions of the planters could only deteriorate since they could not harvest, but also lost all their initial investments, including potato seedlings cautiously set aside during winter, despite enduring hunger all along the cold months of the Bosnian context.

➢ Duration :

Another element of the time factor, already partly discussed in an earlier section, is the duration of the disaster. For instance, in the example of Rwanda; areas have been more or less all affected, but some of them just for a few days while other for a long duration, amounting some times more than half a year. In the first case (most of the North and eastern parts of Rwanda), crops in the field, including roots and tubers, were only marginally affected, leading to a high degree of food and planting material availability. Farmers had to hide only a few days in the *bananeraies* and were quickly back in their *rugos* to take care of their crops. In the other areas, especially the south and centre of the country, waves and waves of displaced people pillaring the crops in the fields and looting whatever food could be looted have been followed by the establishment of camps for internally displaced people. Since they could not really sustain themselves with the food aid (low quality and monotony of the rations), they took an heavy toll on the local food stocks, e.i. belonging to local residents of the neighbouring hills, therefore affecting the overall food security of the areas with its various components (including the overall availability of seeds and planting material, as well as wood and other cooking energy).

➢ Frequency :

The frequency of a crisis of a certain magnitude is at the origin of most of the anti-risk strategies of human groups. In areas where the state of crisis became a norm (sahelian agro-pastoral fringes, etc.) rather than the exception, more complex mechanisms tend to develop. In areas which have been crisis-free for a long period, the population is often less well-equipped to deal with crisis. The degree of development and the sophistication of the anti-risk (prevention) and risk coping (adaptive measures when the disaster finally strikes) strategies and mechanisms are closely linked to the way a whole society (both the collective structure and the individualities its is composed of) perceive the dangers, its magnitude and probability of occurrence.

Analysis of anti-risk strategies in drought-prone areas has shown that there is a direct relation of proportionality between the probability of drought and the number of varieties used per crops. These varieties not only differ in terms of length of their physiological cycle, but also in terms of resistance to the various possible stress. In addition, a few studies of the genetic characteristics of some of these varieties have shown that even the apparently most homogeneous varieties were in fact genetically heterogeneous populations. The higher is the degree of risk, the more numerous are the varieties used. It should be here noted that any analysis of crisis-related phenomenon that would use an evaluative approach through the use of average data will be bound to non-useful and irrelevant results. Indeed, in either climatic or man-made disasters, what counts is the risk, estimated as a probability of occurrence. Only frequency analysis based on the theory of probabilities can be of use. It is therefore also important to try to perceived what are and how are made risk calculation by farmers. This issue of frequency can also be linked to the "duration" factor of a possible disaster. In fact, this link (the probability for a phenomenon not only to occur, but also to last) has been at the origin of most of the traditional food and seed stock strategies.

In Somalia, where climatic uncertainty is more than just an obvious factor but a part of the whole social fabric of the Somali society, storage technologies based on underground pits have enable villages to keep food and seed stocks amounting to more than 5 years of consumption needs and for adequate agricultural activities. This means that the village was able to cope with 5 consecutive years of bad harvest without much harm. In the Sahel, the establishment of collective granaries under the control of the village council had exactly the same purpose: to keep stocks to cope with the "probable unexpected" event such as drought, desert locusts, plant diseases, etc. This can also help to understand the validity of certain indicators and early warning signals used by these populations in their risky environments. This of course applies mainly to drought, flood or pest-related disasters, but experiences in long lasting internal wars (such as Angola or Afghanistan), shows that rural population use the same approach to estimate the risks of attacks.

As much as the form of the clouds or the direction of the wind can inform on climatic events to come, increases in food purchases by soldiers in cantonments have often been associated with soon to come military operations.

3.2. The resilience factor

Crisis and disasters are not an uncommon feature of history. Societies have been able to cope with them or have disappeared. Of course, the above mentioned time factors (timing, duration and frequency) as well as the magnitude and scale of the phenomenon will have specific effects to which these societies and their various components will react more or less appropriately.

- If the disaster is of a limited scale or duration or if it is a relatively frequent feature which has generated adaptive measures, most societies will resist with only little harm. They have developed over time (and time here means decades, if not centuries) adaptive practices where risk is both minimised and spread over years, crops, seasons and ecological conditions. Although some of the adaptive mechanisms can have negative impacts, including increased destitution of specially vulnerable groups or migration, every one thinks the situation will come back to normal relatively rapidly.

- When magnitude, scale or duration increase, changes in feeding patterns can be seen as a recorder, as well as the initial phases of selling of assets. The poorest social strata, which of course has the thinnest buffering capacity (capital and assets that can be lost without affecting the overall survival capacity of the family) will often become more specifically affected.

- Components of the society which are still very close to their natural environment and know and are able to manage these natural resources will be less affected, as time runs on, than groups or populations cut from their natural resource basis. What has been observed in context such as Mozambique is that the more traditional groups (mainly in the RENAMO zones) were less affected than the ones which have been removed from their natural environments and regrouped in artificial settlements (*aldeas communais*). The later were disconnected from most of the wild foods and traditional varieties and, therefore, were more vulnerable to disasters.

- When the situation further degrades, what counts then is daily survival. Survival mechanisms, such as reduction of the diet, relying on wild foods (leaves, forests fruits and tubers, etc.) are the last protection against total destitution. In these contexts, characterised by a high (if not total) degree of scarcity, farmers and non-farming community have often had to rely on unsustainable practices: charcoal making leading to deforestation, exhaustion of reserves of wild food, consumption of lactating she-camels, cows or goats, etc. Among these features, which often comes along with large and protracted disasters and their specific post disaster situations, consumption of the last stocks of seeds is also a result.

Here indeed, the fact that human societies are differentiated has to be central to our analysis. The socio-economic stratification of these societies of course differ from one context to another, but it is this phenomenon which will generate various aptitudes and capacities to cope or not with various level of "aggressions" on the "family units" and "societal systems".

The notion of "vulnerable groups", as imperfect as it is, became a key concept for the aid actors. However, the means and methods to identify them, to establish the roots of their vulnerability and to understand their strategies are still to be developed:

- Who are these vulnerable people?
- Why are they vulnerable?
- Are they a group (with rules and structures) or just a constellation of individuals, each of them in his/her own drama?
- How are they coping with the disaster?
- Are they taken care of by the rest of society or left aside?
- How will they manage to restore a minimum productive capacity after the disaster?
- More specifically, how are they dealing with the "seed issues"?

The tools needed to answer to these questions belong more to the field of social sciences than to agronomic and seed sciences. This explains partly why agronomists with no or little background in social science often have problems understanding what is happening and to relate seed issues with evolving social fabrics of disaster affected societies. A kind of scale of the evolutions for the whole system, for specific groups and, more specifically, to the "seed system" can be drawn up (Table 3).

Table 3. Evolution of the impacts of disasters depending on their magnitude and duration

Magnitude of the disaster	Duration of the disaster	Impact on social structures	Impact on vulnerable groups	Impact on the "seed system"
Low	Short	Nihil.	Limited.	Nihil.
	Long	Increased differentiation.	Tend to restrain themselves to mere self- sufficiency.	Focus on local seeds of the most resistant varieties, including tubers and root crops.
Manageable	Short	Nihil.	Are more relying on external jobs.	Nihil or limited.
	Long	Increased differentiation.	Apparition of unseasonal practices. Changes if feeding patterns.	Search of seeds in the markets or in non-affected neighbouring areas.
High	Short	Limited.	Changes if feeding patterns. Selling of assets.	First signs of seed consumption. Search of seeds in the markets or in non affected neighbouring areas.
	Long	Increased differentiation, Apparition of new types of power structures; first signs of elimination of the most vulnerable.	Changes if feeding patterns. Selling of assets. Increased reliance on small job and selling of manpower.	Increased signs of seed consumption, together with decapitalisation of the productive system (selling of seeds, tools, animals, etc.).
Very High	Short	Apparition of new types of power structures; first signs of elimination of the most vulnerable.	Rapid movements of the population.	Increased consumption of seeds. And related risk of disappearance of certain varieties.
	Long	Destruction of the social system.	Large scale death and starvation. Movement of the population, not immediately followed by a return.	High risk of disappearance of certain varieties.

One will notice that no difference is made here between natural and man-made disasters. Indeed, at this stage, there is no difference. Bullet injuries set aside, a multi-year drought will affect a population in a similar way to a multi-year internal conflict.

3.3. The "global context factor"

The vulnerability of a rural economy at large, of farming systems more specifically and of access to seed and planting materials as final point of focus depends largely on the overall system existing during "non-crisis times".

For long times, Aid providers and Donors were focusing on the developing world. General and specialised programmes for disaster and post-disaster seed provision were concentrating on subsistence farming systems where farmers have usually been the main producer and keeper of their own seeds. These systems were based on a multitude of cultivars and local varieties, with often very poetic names. The relation between the site specificity (soils, micro-climatic conditions, etc.) and the types of seeds is very tight. On that side of the hill, you use this type of sorghum; on this side, this other type. In the lower part of the valley, farmers will use one variety of rice while the fields just above will be planted with another one. Each family tries to keep seeds of each of these varieties, and if necessary, can always find a neighbour who has manage to keep a small surplus of a specific one. It could become available if a neighbouring family needs extra quantities of a specific cultivar. In most instances, this system works. In parts of Rwanda, Afghanistan or Angola, this has been the key to recovery. However, if the disaster is of a high magnitude beyond what the community can cope with or, more damaging, if it lasts too long, then some of these local varieties can be lost forever, unless a small quantity has been gathered and kept in a genebank somewhere.

In recent years, Europe (especially former-Yugoslavia and Albania) and the former Soviet Union have been incorporated into the list of war-affected countries. Their agricultural systems are much more sophisticated, with specialisation between seed production units and food production units. This specialisation can even be regional. In former-Yugoslavia, most of the seed production areas (for corn, potatoes, vegetable and wheat) where in the Save-Danau fertile plains currently shared between Croatia and Serbia. When the war broke out, many areas, such as Bosnia-Herzegovia or the former United Nations Protected Areas (UNPA) of Krajiana were cut off from seed production zones, and therefore, from seed supplies. The same has been observe in the Caucasus region. Most of the potato and wheat seeds were produced in the Krasnodar region north of the Caucasus mountains. When the Soviet Union collapsed and some of the former internal borders became actual frontlines, farmers where left alone without any seed supply. However, loss in varietal choices would become more difficult since these seed production units always keep either samples of their own varieties, or more generally produce only varieties coming from research centres. For instance, corn varieties grown in Northern Krajina were all coming from the same production units in Croatia. With the war between Croatia and Krajina, these seeds could not come anymore from this centre. However, Croatian seed production units used to get their seeds from a well-known agricultural research station in the Voivodina region. This station also used to supply the *agro-combinat* of Seme-Sombor in Serbia. Therefore, the seeds of the same varieties could be found from an alternative source through an alternative route.

Systems based on "one agricultural season per year" (such as the Sahel) are much more sensitive than systems based on two or three cropping seasons (because of long rainy seasons or irrigation facilities). Another very specific factor affecting the "robustness" of a farming system against disaster is the types of crops it is based on. The terms of this robustness are based on a series of factors: mono-cropping versus multiple crops, homogenous length of plant cycle versus utilisation of early, medium and late maturing varieties, cereals versus tubers, hybrids versus open pollinated varieties, etc. In each of these different systems, the main element that characterises the conditions during a disaster is "certainty or uncertainty" of the occurrence of certain events that would have been considered as "ascertained" during non-crisis times.

3.4. The "aid factor"

It should be remembered that, in times of disaster and post disaster, the Aid community (donrs and implementing agencies alike) becomes a crucial actor. With its various programmes (food aid, seed and tools distribution programmes, etc.), it can play a complex role, with positive and negative effects on:

- Overall recovery, at both socio-economic and food security levels;
- Sustainable seed availability (production, supply, etc.); and
- Overall preservation of plant genetic bio-diversity.

Experience has shown that, for the sake of doing good, some times a lot of harm can be done. If logistics, which occupy large part of the time of Aid workers is a relatively simple process, supporting agricultural recovery is complex. NGOs have, in the last five to ten years, become numerous in the field of seed distribution. But due to the dominance of logistics-oriented, unprofessional or unprepared staff, a lot of mistakes have been committed. This is especially unfortunate since the process of "greening the ruins" is often carrying with it a lot of hope. There is nothing as sad as a farmer who spent his last resources and energy to plant fields that will yield no harvest.

After decades when the only answer from the International Community has been "food aid", a new feature of relief programmes became more and more frequent - the tendency to, at least theoretically, "link relief with development". This has been on most agency's agenda since at least five years. This has taken various types of forms, from mere rhetoric (developmental relief) to highly sophisticated recovery programmes. Among these programmes, seed distribution became fashionable within both the humanitarian agencies and the donors. These types of activities were indeed supposed to both limit the scale and duration (and therefore the cost) of food aid programmes (by making people more quickly self-sufficient) and bridging between pure relief and rehabilitation/development programmes. This approach, based on the concept of "*continuum*" described earlier, suffered a certain number of setbacks. In certain countries, such as Angola, free seed distributions have been organised for more than 10 years without establishing even a minimum of sustainable food security. In countries like Somalia, war affected areas with high scale of destruction and sufferings (including famine) are bordered by peaceful zones, where the full fledge agriculture cycles are still at work and even able to produce surplus. This, which led to the new paradigm of "*contiguum*", will bear a heavy weight in the chance of preserving local phytogenetic resources and accessing locally known varieties.

Often, diagnosis has been of a limited quality and the approach more of a "pizza delivery" nature. Unappropriate varieties have been purchased. Seeds of a special type were brought to the wrong areas. Delays in the order and delivery brought the seeds too late in the hands of the farmers. The frequent absence of the required "seed protection food ration" led in some instances in high seed consumption rates. Many factors can result into a low performance of the agricultural recovery programmes.

Three key issues have to be given the proper attention:

- Food aid versus assistance for agricultural recovery;
- Imported seeds versus local purchases, and
- Free distribution versus provision on a loan basis.

Although these points will be treated in other papers, it is worth mentioning them here since they shall be given the appropriate consideration in any strategic preparedness plan of action.

4. CROSSING FACTORS: A PRAGMATIC TYPOLOGY

As the diversity of disaster situations has been established, it is then useful to organise this knowledge in an orderly manner. A crucial points is that the same disaster does not affect the same way different types of contexts. Similarly, each of the farming systems can be more or less resilient depending on the type of disaster. On the basis of the key variables related to a simplified classification of disasters and to existing agrarian systems, the following double entry grid has been elaborated. Then one can place in the grid known past or ongoing disasters - a few examples have been positioned in this type of grid in Table 4.

Although this grid is by no way a "miracle recipe", it can help to reference and organise the knowledge about disasters and their impact on farming systems. This approach, therefore, can assist future strategic planning exercices.

5. DISASTERS: GAPS IN INFORMATION FLOWS TO BE FILLED UP

Early Warning Systems have existed now for many years and are becoming more sophisticate. The most sophisticated and reliable ones are, up to now, the ones targeting specifically climatic disasters (droughts, floods, etc.) as well as those with relevance with locust infestations. Early warning signals

Table 4. Grid crossing types of disasters and types of affected contexts

Characterising disasters / Types of agrarian system				Manual subsistence agriculture where farmers produce and keep their seeds.	Intermediary stages with various level and degrees of dependency on seed supplies external to the production unit.		Highly mechanised systems where farmers buy all their seeds from specialised companies.
Types of disasters		**Onset**	**Duration**				
Natural disasters		Slow onset	Short duration				North Korea since 96
			Long duration	Drought in Sahel in 73-74			
		Rapid onset, unpredictable and unexpected	Short duration		Floods in China in 98		Landslides in Italy in 98
			Long duration				
		Rapid onset and predictable event	Short duration	Locust infestation in East Africa		El Nino in 97-98 in South America and Asia	
			Long duration				
			Short duration				
			Long duration				
Man- made disaster	International war	Slow onset	Short duration				
			Long duration				Former-Yugoslavia
		Rapid onset, unpredictable and unexpected	Short duration	War between Senegal and Mauritania			
			Long duration				
		Rapid onset and predictable event	Short duration				
			Long duration				
		Other	Short duration				
			Long duration				
	Internal conflict	Slow onset	Short duration				
			Long duration	Uganda since 96		Palestine	Azerbaijan/ Armenia
		Rapid onset, unpredictable and unexpected	Short duration	Erithrea/Ethiopia war			
			Long duration	Rwanda, Burundi			Afghanistan
		Rapid onset and predictable event	Short duration	War in Zaïre 96-97			Albania
			Long duration	South Sudan			Kosovo
		Other	Short duration				
			Long duration				
	High tension leading to socio-economic disturbances	Slow onset	Short duration			Indoneasia	
			Long duration				
		Rapid onset, unpredictable and unexpected	Short duration	Guinea-Bissau			
			Long duration				
		Rapid onset and predictable event	Short duration				
			Long duration	RDC			
			Short duration				
			Long duration				

related to social upheavals, internal conflicts or international wars also exist. They are, in fact, the basis of intelligence and diplomatic activities since centuries.

However, most of the early warnings issued in the last decades have not been taken seriously, or responded to, by the international community. In many instances, they were totally neglected. More sadly, some of them have been purposely not listen at. Reports on the developing famine in Ethiopia were available and circulated months and months before the situation turned into the most well-known famine of the mid 80's (the WFP/FAO donor conference on Ethiopia had been held without any response until the BBC footages showed the developing horrors at Christmas 1994). When early warnings are related to ethnic or social crisis, they have either not been noticed (Rwanda, in the early nineties a few years before the genocide actually took place) or been followed by inappropriate measures (Kosovo, which was raising a lot of concerns from the "educated observers" since mid-80s was not included in the Dayton agreements). The logical consequence of an early warning, which is a rapid response, has in most instances not materialised.

However, relatively simple cost effective measures could have been taken in terms of setting aside local seeds, and increasing local, national or regional areas under seed production. More important, strategies for *in situ* protective measures on the bio-diversity could have been implemented in a certain number of cases. In fact, most of these measures should become full fledge parts of any agricultural strategy aimed at promoting sustainable development.

Available information systems relating to disaster situations are plentiful. Yet they are not used properly, i.e. they are rarely listen to. Most of the information required is available, but often political priorities prevent a proper use of them. There is, nevertheless, room for improvements, at least in the way pre-crisis information can be collected and used before the disaster strikes.

6. DISASTER PREPAREDNESS: THE FORGOTTEN PRIORITY

The problem with preparedness measures is that they are an investment which will only pay off if and when the disaster actually strikes. This explains why neither local authorities nor Donors are very keen to finance them. Unfortunately, when disaster strikes, it is often too late. Then costs run high, appropriateness of the response is often limited and suffering reachs peaks that could have been avoided.

There is, however, a lot of confusion behind the term "preparedness". It is often understood only from a logistical angle. Pre-stocking of goods (and in relation to this paper - seeds) and transportation means is of course important. However, preparedness is more a state of mind than a physical process. With this overall acceptation of the concept, preparedness has to do with a few key strategic factors which include, but are not limited to :

<u>Acquiring the knowledge:</u>

In many man-made disasters, there will be no Ministry of Agriculture to deal with, no technical decision makers available for discussion and agricultural recovery. Farmers could be the only source of information. The thoroughness, but also the limitation of this information (site specific, little global view, limited knowledge on either the world seed system) the absence or excess of knowledge of the humanitarian business can lead to very biased assessement. By chance, there is no region on Earth which has not been studied, especially disaster prone areas. How can this information be collected, compiled and made available before the crisis, so that when it occurs, useful information is immediately accessible? Information systems connected to the INTERNET are one of the most recent assets of the international community. More and more, people's groups, NGO, local institutions can get "on the net" to look for required pieces of knowledge they might need. Structures like FAO/GIEWS or specialised centres can insure that this information is available in a "ready to use" form. In addition, trusted and tested solutions often do exist. What counts is the availability of the information about them. Local seed banks have been set up in Western Africa and in India. Trial and error programmes on that subject have already been in place in various areas of the world. There is no need to repeat the same errors, but avoidance of doing so depends largely on whether or not the actors are aware of these lessons, and make use of these lessons learned.

Be ready:

The establishment of seed stocks or lists of useful appropriate seed varieties and seed supply sources requires competent, dedicated and convinced staff. The human resource factor has often been seen as one limiting very much the efficiency of programmes. Training on matters related to seed issues with hands-off sessions for staff already working and specialised modules included into agricultural schools and universities curriculum can easily be organised.

Developing inventories which include a listing of local varieties with their characteristics and site specific adaptability will be crucial. One has to remember that *in situ* varietial conservation is often more efficient and cost effective that the establishment of costly national or international *ex situ* conservation systems: national seed banks require not only a sophisticated infrastructure of cold rooms, refrigerators and packaging units, but also agricultural stations for regular regeneration of the seed stocks. These national systems are, however, still necessary, since they could permit national inventories and co-ordination of overall biodiversity conservation.

International systems have also shown their usefulness in certain contexts. Most of the rice varieties of Cambodia were stored at IRRI. During the Khmer Rouge era (1975-78), some important varieties where almost lost (especially floating rice brands). In the mid 80's, IRRI managed to bring back some of these cultivars which then could be re-multiplied. In Angola, the national collection of sorghum, maize and beans was located in the premises of the Huambo National University of Agriculture. The military occupation of this town during the civil war and the destruction of the compound of the university resulted in the destruction of the collection. If some samples of the Angola varieties would have been provided to the regional phytogenetic resources conservation programme, the losses would not have been so detrimental. Now, not only the collection has to be started from scratch, but it is not sure if all of the varieties will ever be recovered. The decades of famines and population movements have taken a high toll on the national phytogenetic heritage.

Establishment of village based seed banks, with the necessary training and social work in order to sensitise the population of the importance for them to keep these banks in good conditions and to use and resupply the stocks is needed, and requires appropriate approaches.

These preparedness measures are especially important when a disaster has been followed by large scale displacements of rural populations. Indeed, when the refugees or internally displaced people go back to their areas of origin, appropriate types of seeds are often not available. Since most of these events (e.i. return and repatriation) are often predictable (this has been the case in most instances, apart of the recent massive return of Rwandan refugees from former Zaire), appropriate establishment of stocks of local seeds could have taken place. In Cambodi, for instance, the repatriation preparation process took more than a year. Alas, only very few people mentioned beforehand the importance of pre-stoking of Khmer traditional rice varieties. When the refugees came back, they only received "modern varieties" often very vulnerable to flood or drought. It did not help the economic re-integration of these returnees nor the establishment of an acceptable level of food security. This is the kind of failure which underlines how weak the international community's preparedness response has been in recent times.

Think ahead:

With a bit of proactive thinking, many steps can be achieved long before the emergency.

During the post conflict or post disaster period, access to traditional varieties will be crucial to re-establish farmer seed stocks and restore risk minimisation mechanisms based on socially-owned knowledge. Pre-crisis establishment of national seed banks and/or bio-diversity genebanks will be crucial for the post-disaster recovery. Regional co-operation is needed. The connection of national strategies and perspectives with regional ones will help to distribute the risk and share costs. Experience in Angola has shown that this can be done in a cost effective and efficient manner, provided that certain means are available and matched with a real will of local actors.

We also know that, in post-disaster periods, the absence of law and order can facilitate all kind of pillage. Seed companies can jump to collect samples of seeds they will use in their own research work without even mentioning the origin of the genetic material. Will the "protective components" enshrined in the Convention for the Protection of Biodiversity shall be actively defended in post disasters contexts? After all, countries back from the brink should as well be respected and their genetic resources protected.

7. CONCLUSION

Among the many factors contributing to the success of humanitarian disaster response operation in general and of sustainable agricultural recovery through seed distribution, one of the most important is the quality of the diagnosis.

This diagnosis depends, among other factors, primarily on:

- A proper understanding of the nature of the disaster, of its characteristics;
- A thorough analysis of the system that prevailed before the disaster took place, including the production processes (in terms of timing, requirements, relations with other sectors, etc.);
- An in-depth assessment of the way the disaster has affected life in the rural areas, including the agricultural cycles; and
- An identification and appraisal of the adaptive mechanisms and survival strategies put in place by the affected populations.

8. ADDITIONAL READING:

AUBERTIN, C. 1998. Les enjeux de la biodiversisté. Poche Economica, Paris, 112 p.

BESSET, J.P. 1992. René Dumont: une vie saisie par l'écologie, Collection"au Vif", Stock, Paris, 375 p.

CARBONNIER, G. 1998. Conflict, post war rebuilding and the economy: a critical review of the literature. The war torn societies project. UNRISD, Geneva, 76 p.

FAO. 1996. Conservation and sustainable use of phytogenetic resources for food and agriculture. FAO documents prepared for the World Food Summit, Rome, Italy.

FAO. 1997. Contributions for the development of seed security strategies in Disaster-prone Regions. Proceedings of the Florence International Workshop on seed security for food security, 231 p.

FAO. 1998. La production des semences de qualité déclarée au Rwanda; Kigali/Rome, 83 p.

GARY, J.M. 1995. Ethnobotany: a people and plants' conservation manual. WWF/UNESCO, Chapman & Hall, 268 p.

GRALL, J. and LEVY, B.R. 1985. La Guerre des semences: quelles moissons, quelles sociétés? Ed Fayard, Paris, 410 p.

GRUNEWALD, F. 1985. Le riz et le sucre: manuel methodologique pour la recherche développement en agriculture dans le sud-Thaïlande. GRET, Paris, 26 p.

GRUNEWALD, F. 1986. *Alarme précoce, aide d'urgence et développement, quelques idées pour une stratégie*. Bangkok, Asian Disaster Prepardness Center, 1986, 57 p.

GRUNEWALD, F. 1989. "Stratégies paysannes à la rescousse de l'autosuffisance". In: *Affaires Cambodgiennes.* Paris, L'Harmattan, ; pp.149-174

GRUNEWALD, F. 1993. "Pêche, forêt et agriculture dans un Cambodge à peine sorti de la guerre civile". In: Revue Tiers Monde, XXXIV, N134.

GRUNEWALD, F. 1993. When the rain will return: emergency assistance, rehabilitation and development. Text presented at the Conference Colloque "Développement, il y a urgence, Lyon", November 1993, 13 p.

GRUNEWALD, F. 1995. "Action before, during and after the crisis: the experience of ICRC in retrospect". In: International Review of the Red Cross; N° 306, May-June 1995; pp. 263-281

GRUNEWALD, F. 1995. "Protection of biodiversity in times of war and during post conflict periods". Report of the first International Conference, Huambo (Angola) July 1995, ICRC, 25 p.

GRUNEWALD, F. 1996. "The other cost of war: protection biodiversity in times of conflict". In: *Plant Talk : plant conservation worldwide,* N° 822, Novembre-Décembre 1996; pp. 633-654

GRUNEWALD, F. 1996. "For or against food aid". In: *International Review of the red Cross*; N° 312, October 1996; p. 15

GRUNEWALD, F. 1997. Au delà de la survie: concepts et pratiques de la réhabilitation agricole au CICR. "Working papers of the Relief Division", Genève, 27 p.

MACRAE, J. and BRADBURY, M. 1998. Aid in the twilight zone: a critical analysis of humanitarian-developement aid linkages in situations of chronic instability. UNICEF, New York, NY. 75 p.

MAZOYER, M. et al. 1998. Histoire des agricultures du monde: du néolitique à la crise contemporaire. Seuil, Paris, 531 p.

ODI. 1997. Fournir des semences pendant et après les situations d'urgence. Programme de semence et de biodiversité. Revue de l'Etat des connaissances N°4, London, 151 p

ORSTOM. 1989. Le risque en Agriculture. Collection: *A travers champs*, ORSTOM, Paris, 619 p.

SPERLING, L. 1996. Rwanda: Household survey on status of potato crops. Seeds of Hope, ISARD, Kigali, 27 p.

SPERLING, L. 1996. Rwanda: Nation Household survey on status of beans, sorghum and Cassava crops, 1995: the impact of war and varietal erosion, Seeds of Hope, ISARD, Kigali, 60 p.

SPERLING, L. 1996. Executive summary and reflection of the Seeds of Hope socio-economic analysis in Rwanda: the impact of war on agricultural production. Seeds of Hope, ISARD, Kigali, Rwanda. 24 p.

USAID/DESFIL. 1996. Seeds for disaster mitigation and recovery in the Greater Horn of Africa. Chemonics International, Inc. Washington, D.C. 137 p.

FARMER SEED SYSTEMS

Temba M. Musa
Borrowdale, Harare. Zimbabwe.

1. INTRODUCTION

1.1 Background

"Farmers' seed production" essentially refers to growing a crop of which part is saved as seed for own use. Saving the best grains, roots or tubers from consumption, their storage and planting developed over centuries into structured local seed systems. Therefore, seed is the primary agricultural input, and is the embodiment of the farmer's future harvest. This system is variously described as 'farmer seed system' or 'local seed system'. The term 'informal seed provision[11] system' essentially refers to the same, emphasizing that the farmers' seed production takes place outside the formal regulated seed production sector. Such systems, which are directly linked to the adaptation of the farmers to their agro-ecological and cultural environment and support farming systems based on number of crops and varieties needed for household and on-farm use are the main agricultural resource for resource-poor small scale farmers in developing countries.

Developing countries constitute approximately two thirds of the world's agriculturally productive land area and support about three quarters of its population. Because of the inherent resource limitations, these areas include some of the poorest segments of the world's population who battle for their livelihood from small scale farming systems. It is obvious that any conditions which can disrupt small scale farming systems, may have devastating long-term implications for household, if not national, food security as this often also disrupts farmer seed systems. However, in the past decade some of these areas have experienced a propensity for devastating natural disasters and/or armed conflagrations. Because of this, rural households in these areas are food deficit in most years, and poverty limits their capacity for additional food purchases. For this reason most rural households have to put priority on obtaining food and incomes for the present, which leads to over-taxing of national and international relief resources. While small scale farmers often cannot afford to take a long-term view in regard to resource conservation, an international crisis is looming as donor fatigue has set-in resulting in a decline in funding for disaster mitigation relief efforts world-wide.

There is, therefore, a pressing need for strategies that will improve household food security and incomes in these areas, while at the same time creating conditions conducive for sustainable self-reliant seed production in affected areas. The prevailing mood among national and international aid agencies is that, despite inevitable problems, support of imperfect farmer seed systems may be a more sustainable disaster mitigation strategy, than supporting imperfect and often dysfunctional state managed seed relief enterprises in these areas.

1.2 Achievements

That informal seed provision systems are resilient to environmental changes, which at times are catastrophic, and agricultural advances is evidenced by the observation that they still contribute an estimated 90 percent of all seed used for food production in developing countries (Almekinders *et al.*, 1994). Even in highly advanced agricultural systems found in most countries of the North, farmer-

[11] Seed Provision Systems in the context of this paper refers to all processes related to variety selection, breeding and development, seed crop production and harvesting, seed conditioning, processing, storage, marketing distribution and technology diffusion

saved seed contributes to over 30 percent of all seed for self pollinated food crops (Ghijsen, 1996; Jaffee and Srivastava, 1994).

This paper explores how limited-resource farming households and communities presently manage their local seed provision systems, in particular emphasizing on their capacity to survive under recurrent emergencies. Further exploring the interaction between these systems and other institutions such as, National Agricultural Research Systems (NARS), International Agricultural Research Centres (IARCs), Non-Governmental Organizations (NGOs), etc., in developing and adapting them to meet new challenges. Such a critique will enhance the establishment of international programmes designed to generate and exploit spillover benefits from these experiences.

1.3 Issues for consideration

The paper is divided into three parts. The first part will review the socio-cultural and economic issues that determine demand for seed, and draws implication for possible organization for seed supply systems to satisfy household food security. The main focus will be to outline the stages in the development of local seed systems and sketch out the roles and responsibilities of the various components that contribute to each stage. The second part of the paper presents an analysis of the strengths and weaknesses of local seed systems. It focuses on competencies and incentives of each component involved in trying to devise strategies for strengthening local seed provision systems based on transaction costs that characterize collaboration among different types of organizations. The final section presents some possible intervention mechanisms which may be effective in quickly identifying and alleviating the effects of a disaster in limited -resources communities.

2. FARMER SEED PROVISION SYSTEMS

2.1 Diversity in an integrated system

Farmer seed provision systems are described as systems in which selection, seed production and seed diffusion or exchange are integrated with crop production and are defined in relation to the farming system, of which they are part of. The components of such farming systems being the farmer, the agro-ecological, socio-economic and cultural environment.

Although each farmer seed system tends to be location specific in terms of the combination of crops, varieties, seed, farmers and environment, this section will present a typical characterization of such seed system.

2.1.1 The seed system as part of a farming system

The typical small scale farm in developing countries combines a wide range of crops and varieties to meet the diverse objectives. The rationale for this are is that apart from household food security, this diversity is required to meet the demand for social amenities, health and socio-economic needs. For most crops and varieties, each individual household meets its own seed requirements. Seeds, which in the context of most farmers refers to any plant part which can reproduce the same kind of crop, are selected from crop production, in some cases the selection process can be elaborate. However, under very special situations where certain seed qualities, especially those related to variety genetic maintenance or improvement, are desired seed produced in separate plots. Experimentation is another important aspect of the seed production processes related to farmer seed systems

2.1.2 The seed system as part of farmer knowledge.

Farmer seed systems through repeated crop production and selection under local conditions, allowing for the effects of mutation, seed exchange, hybridization and selection pressure exerted by both natural factors and humans, represents a dynamic evolutionary system. Knowledge to manipulate plants to suit farming systems is well developed, as may be evident from the development of crop varieties out of wild species. This knowledge has been accumulated over centuries and generations of farming in the same, or similar, environment and has contributed immensely to the resilience of these seed systems to periodic aberrations in the crop production conditions which at times impact negatively on farming systems.

Farmer knowledge is not always apparent, as it is often influenced by socio-cultural parameters not easily comprehensible to most researchers and often seemingly incongruous with modern 'scientific' type of crop development. Brush *et al.* (1982) described a form of folk-taxonomy of the potato varieties based on agronomic characteristics such as skin colour, processing and culinary quality. Such folk-taxonomy are important for farmer variety selection and are a common feature for most farmer seed systems (Musa and van der Mheen-Sluijer, 1997) and have been used successfully in some breeding programmes (Monyo, 1998).

These two categories which define the characteristics of farmer seed systems while by no means comprehensive, serve to identify the various some selection criteria critical to the survival of these seed systems. These include:

(i) Feasibility under given socio-economic circumstances.

There is evidence that farmers capacities as breeders-selectors is extensive based on the level of adaptation of their varieties to local conditions. The grid selection is often accompanied by 'shuttle-breeding', which involves a process for shifting the selected plant populations between two different fertility environments The skills for grid-selections have been reported for chickpea adaptation for a particular environment in Ethiopia (Berg, 1994) and sorghum and pearl millet selection to resistance to Striga sp. in Ethiopia and Zimbabwe (Berg, 1994; Musa and van der Mheen-Sluijer, 1997). This skill is, however, limited to a few local farmer seed specialists, normally with larger land-holding than the average small scale farmer.

There is, however, little systematic information on similar occurrences for a general extrapolation to other regions or crops to be made, suffice to say variants of it may exist.

(ii) Correspondence with farmer's goals and preferences.

Many crops have multiple uses. Maize and barley produce grain for consumption and leaves and stalks for livestock-feed. Some sweet potatoes, cassava and cowpea varieties are preferred because their leaves make good vegetables, while the tubers, roots and grain can be processed for food fresh or dried. In most cereal based farming systems, plant and animal production are integrated. Straw yield and straw quality may, in many cases, be at least as important as grain yield. Short straw varieties may thus not become very popular, where they are used for animal feed or for building materials (Musa and van der Mheen-Sluijer, 1997).

(iii) Feasibility under given natural conditions.

Different crops and varieties also provide an opportunity to use genetic diversity for optimal exploitation of available resources. The different planting and harvest dates of crops spreads the labour demands for ploughing, planting, weeding and harvesting. Two matching crops in mixed cropping make better use of water, the minerals and the space than when planted separately. Another important additional advantage of mixed crops and varieties is a reduced insect and disease pressure.

In general, the same principle which applies to mixed cropping and variety mixes apply to heterogeneous landrace varieties (farmer traditional varieties). Only because two genotypes of pearl millet look more alike than, for example, a cowpea and maize plant, the advantages of mixing will be less clear and more difficult to measure. However, farmer seed systems favour this type of varietal and crop mixes.

(iv) Ecological viability.

Different crops and varieties provide an opportunity to make better use of available resources. Longley (1993) described how labour-constrained Mende women rice farmers have developed different varieties to exploit variation in vegetation, land topography and changes in soil moisture regimes obtaining in their areas to successfully provide food throughout the year based on one crop. As rice is largely a self pollinating crop, subject only to small amounts of cross pollination, women resorted to individual hand panicle harvesting within a crop. This gave these farmers the option to reject 'off-types' or collect such 'off-types' for further experimentation. These farmers practiced a form of mass selection through this processes of selecting-out off-types in the course of panicle harvesting. Further, such action stabilized the main seed types, while effectively bringing about a systematic grouping among the off-types when planted out.

This form of panicle selection combined with shuttle-variety selection[12] (Berg, 1994), has resulted in the differentiation of Mende rice germplasm into three distinct ecotypes. Short-duration rice ecotypes which ripen in three to four months are planted on moisture retentive soils at the foot of the valley slopes. Medium-duration higher yielding varieties with longer and stronger rooting systems, maturing between four to five months, are planted on free-draining upland soils when the rains have well set-in and provide the bulk of the grain harvest to this day. Long-duration types, ripening between five and six months and adapted to variable flooding levels (including floating types) are planted in the beds of river courses and inland swamps - they are often planted during breaks between operations for establishing upland rice. These ecotypes allow the family to stagger their crop establishment, thus spreading the labour throughout the year in the food crop production processes. These characteristics are contained in the seed developed in seed provision systems tailored to meet the complexity in the cropping environment.

These practices allows harvesting to be staggered. Such a farming systems permits the farmer to be food secure without worrying too much about long term food storage. Similar variety diversity has been reported with women field bean growers in Malawi and Rwanda, root and tuber farmers in the Andean regions of South America and maize farmers in Mexico (Lourette and Smale, 1996; Eyzaguirree and Iwanaga, 1996; Musa and van der Mheen-Sluijer, 1997). All these farming systems rely on the explicit knowledge of utilizing varietal diversity maintained in the seed for household food security. However, food security in heterogeneous environments, can also be based on having access to crop genetic diversity, as observed in Mozambique. The knowledge farmers have on cereals, root and tuber, legume and pulse crops has been preserved in their seed systems to meet their household food security at different times of the year. The hunger period between each crop harvest, often is met by fleshy fruits like papaya or banana crops.

The diversity of crops and varieties spreads the labour demands for ploughing, planting, weeding and harvesting. This reduces the transaction costs (labour) for such farming systems under women control, involving limited long-term food storage. This demands a very sound knowledge of the seed to be used in such mixed cropping farming systems, often resulting in farmer seed security to be as complex as the environments they are designed to serve. Farmers make use of genetic variation as a tool in matching environmental variation, rather than make the environment uniform. The best documented case which illustrates the interface between farmer selection and environmental variations in both time and space is that of the Mende rice farmers in Sierra Leone.

---

[12] Shuttle-variety selection in the context of this paper refers to a process whereby a variety is shifted between different environments until its genotype adapts to given environmental conditions.

West African rice farmers have for decades harvested seed from the centre of the field to maintain their genetic purity and avoid introgression from other landraces and other wild *Oryza* species surrounding the field. Sorghum farmers select heads for seed from a range of plant types, to maintain a near consistent heterogeneity in a variety. This is done to maintain the taste and malting characteristics required for various foods, including brewing traditional beer.

The diversity of crops and varieties serves the stability of farm production. Therefore, crop varieties or crop combination that produce each season a reasonable yield may be a better option for resource poor farmers than a single crop or variety with a high average yield, but with a higher risk of producing less than the required minimum to meet household food security.

These categories serve to show that local seed provision consists of basically three components: selection, production and distribution. However, in the main seed production, as part of the general food crop production, is the focal point of this system. This ensures equitable distribution of seed to all those prepared to grow crops, with a potential household seed secure through self-reliance in crop production. While selection, breeding, and crop or variety development are considered important these remain specialised areas, as the demand for new genotypes is cyclic phenomenon, occurring once after one or two decades; for most households such demand is often met through trade.

2.2. Household seed security

The objective of farmer seed systems is to make sufficient quantities of seed for the preferred crops and varieties of optimum quality available for each farming unit every planting season. This broadly defines seed security, recognising that it is both a function of two factors, namely seed availability and seed access. Seed availability generally refers to the amount of seed harvested during the course of food crop production in a farmer's field, and is associated with the following supply parameters: quantity, timing and sustainability. Whereas the access aspect is related to the following parameters: quality (desired variety with the right genetic and sanitary criteria for producing a crop), equity (reaching all farmers) and finance (ability to have from external sources).

The two components important to seed security are generally based on needs and expectations of the farming system interfacing with individual household and community economic and cultural requirements. These basically divided into food consumption, medicinal, construction and clothing and the surplus for market and entertainment. These issues at the farmer level are not gender neutral and their consideration may offer critical insights into developing strategies for mitigating the effects of natural and man-made disasters.

2.2.1 Seed availability component: quantity, timing and sustainability aspects

Seed quantity is basically a function of crop production. Culturally because food crop production and processing is a woman's job, this aspect of the seed supply system has traditionally been associated with women in the informal farmer seed systems. The women generally over centuries have develop an intrinsic ability to select the correct mix of crops and varieties to meet the family's expectations. Since the required quantities of seed for producing most crops is relatively small, women have generally developed the knowledge and skills associated with the conditioning, processing and storage of such seed.

Both the timely supply of seed for crop production and sustainability of the seed system is dependent on the efficacy of the conditioning and storage of planting material by the women. If the harvest is timely, and understanding of ecological parameters for food production is comprehensive (Longley, 1993), then timely supply of seed to suit any environment is possible. For instance, Kpelle women in Liberia maintain over 100 varieties of rice in swidden cultivation, while in the SADC region women have been reported to store seed of ten ecotypes of sorghum and pearl millet at a given period in their homestead granaries (Musa, 1995; Musa and Rusike, 1997; Rohrbach, 1997). Young girls become enculturated into their gender roles as managers of crop and seed diversity, and memorize information about the varieties and the micro-environments to which particular varieties are suited.

In all respects, women are better selectors of varieties and crops to meet all the expectations of the community or household. Seed production for household food security is generally a function of the food crop production, from which seed is selected either during the growing crop or after harvest. Therefore, seed availability is linked to sufficiency of the harvest to last from harvest to harvest, or is a function of household food security.

2.2.2. Access to seed: quality, equity and financial aspects

Crop productivity is dependent on three factors associated with seed: its genetic purity, sanitary purity, physical and physiological purity.

While the crop can not be better than the genetic potential of the variety, the other factors, if improved, can enhance seed germinability and seedling vigour and are a function of seed crop harvesting, seed conditioning, processing and storage. The latter three are closely related to responsibilities associated with women. The former, while a function of the selection process often involves several processes which involved shuttle-selection, breeding and development. Shuttle-selection requires access to land, which often falls in the domain of men in any farming system.

Therefore, access to improved germplasm at local seed provision level is under the socio-economic influences. The men determine the economic importance of acquiring new germplasm either through direct purchase, or providing land for such an exercise for crop improvement. While women have the knowledge for both selection and production of seed crops, the ultimate responsibility for such activities is in the male domain at both household and community level. Examples of various systems employed in gaining access to new germplasm are adequately addressed by Musa and van der Mheen-Sluijer (1997) and Tripp (1997).

Due to inequitable land distribution, most resource poor farmers are unable to grow the desired diversity of crops for sustainable seed security. The most vulnerable sectors of the population has always been women-headed family units. These often become seed insecure because they normally have to eat their stocks of food grain (Rohrbach, 1997; Sperling *et al.*, 1993).

The main problem with local seed systems is that farmers often use poor quality grain as seed, even though seed of acceptable quality may be available, but the farmers often do not have the capacity or willingness to acquire it. Seed security, in particular access to technology, has always been the single most critical factor for household food insecurity.

3. THE NEED TO IMPROVE FARMER SEED SYSTEMS

While farmer seed systems have in the past proved resilient to both natural and socio-political catastrophes due to their intrinsic characteristics of being individually farmer based and heterogeneous in space and flexible in time (Musa and Rusike, 1997; Rohrbach, 1997), the speed at which such catastrophes occur and the population explosion threaten their viability.

Any analysis of the various options for informal local seed provision systems must account for difficulties in categorizing the potential participants, the complex nature of seed provision, and the multiple of sources of seed demand (Tripp, 1997). At the minimum, it must be recognized that the resilience of farmer seed systems are primarily based on their innovative and dynamic nature, which require coordination among different types of organizations. Collaboration between the local farming systems communities, the public, private and non-governmental organization sectors currently play an important role in the development of local seed provision systems. Therefore, for these systems to effectively address the concerns of the farmers they purport to represent, it is necessary to be more specific in assigning particular responsibilities for performance and communication; in short identify the fault lines.

The components of farmer seed systems: selection, development, production, storage and distribution in most cases are part of common crop production processes and as such organised in a horizontal manner. Such seed systems are characterised by a tendency to maintain a wide diversity within and across crops and varieties, or landraces, as a response to diverse ecosystems and local markets. The focal point of farmer seed systems is the crop production phase, which is common to all farmers and thus important to seed security. Each component of the local seed system will be analyzed based on the strengths and weaknesses of the three key aspects critical for their sustainability:

- Organization and ownership of the resources needed;
- Stewardship of the system; an
- Incentives for the performance and technology/information transfer.

The viability of any seed supply option depends on the efficiency and incentives of each individual component.

3.1 Technology transfer

Integrated participatory approaches to seed provision through International Agricultural Research Centres (IARCs), National Agricultural Research Systems (NARS) and public extension services, and NGOs is required. If the system is to be sustainable and self-reliant it will have to have as its main objective to develop technology easily transferable to the target farmer groups. Such technology can only be based on open pollinated varieties which farmers can easily replicate.

3.2 Seed production

Local seed systems usually rely on individual farmers to multiply seed for home use and is often an integral part of crop production. However, it is not all farmers who are capable of producing a crop of the right quality for seed. Seed may be difficult to produce in some environments such as potato and cassava, groundnut and cowpea or beans due to viral or other diseases, and other agronomic difficulties encountered in a crop production system. Seed availability at household level is often, but wrongly included in statistics of seed availability at community level (Musa and van der Mheen-Sluijer 1997).

Communities which have successfully achieved seed security have rather relied on local informal seed producers to produce and multiply seed. In each community, there are recognized seed producers who through their crop agronomic knowledge, access to information are capable of maintaining the diversity in crops and varieties required for local household seed security. Roughly 20 percent of small scale farmers within a community in Zimbabwe provide seed into the local market and have emerged as recognized local seed sources (Rohrbach, 1997). This figure is not significantly different in other parts of the world, in particular Colombia, India and Nigeria, where local seed systems are critical to community seed security (Venkatesan, 1995).

Such farmers have offered the basis for developing sustainable seed supply systems in local communities in both the Indian sub-continent and Sub-Saharan African countries (Musa and van der Mheen-Sluijer, 1997). However, these farmers need training in quality seed production practices and marketing.

3.3 Plant variety selection, breeding and development

The local seed systems are increasingly being forced to depend on public breeding institutions, especially for open pollinated crops. However, public plant breeders' priorities have often failed to address the concerns of the resource-poor farmers as illustrated in the case of Mahsuri rice (Salazar, 1994). The marginal, heterogeneous and remote environments are difficult to address by centralized formal breeding programmes typical of most IARCs and NARS.

Farmer participation in variety testing, selection and breeding can be linked to seed provision which is locality specific (Eyzaguirre and Iwanaga, 1996). The limitations of the centralized breeding programmes, which have been the major constraint to equitable distribution of technology at local level for centuries, have stimulated the exploration of the idea to transfer the final phase of the breeding programmed to the farmers themselves. Instead of the breeder selecting and allowing the farmer to adopt, or reject, farmers are best involved at various stages of a variety development. They may select from a larger number of options, sometimes unfinished products which are not yet genetically stable, this being an adaptation of the evolution of local seed systems based on shuttle-variety development discussed earlier.

This reduces the possibility that the breeder eliminates materials in earlier cycles of selection that are potentially good for particular conditions. It also provides the farmer with a wider range of materials to select from. The promising materials being transferred presumably to cover a wider range of materials, containing more genetic diversity.

- This offers more opportunities for developing local adaptation: final selection is done *in situ*. The genetically diverse populations will evolve differently in different environments.
- An additional advantage is that promising materials move more quickly into farmers fields. Instead of carrying out the 7 or 10 seasons of selection on-station, the breeder may give the farmers materials from much earlier selection cycles.

There are examples of successful landrace enhancement in rice (MASIPAG-Philippines) combines university researchers and local women farmers in the selection, improvement and distribution of local rice varieties (Salazar, 1992), in barley and wheat Ethiopia (Berg, 1992) sweet potato, cassava, pulses and oilseed crops in Mozambique and Angola (Chapman *et al.*, 1997).

Participatory Variety Selection (PVS) is the term for the type of activities in which farmers evaluate and select released or pre-released or advanced (i.e. nearly-finished) varieties. Farmers may be invited to come to the experimental station or to district variety trials to select and take seed from the varieties they consider promising. In another form, a group of farmers may implement a community evaluation trial or farmers may be given materials to grow and evaluate on their own farm.

The greatest draw-back with such projects is that they require considerable input from NGOs whose commitment to long-term projects is still to be fully analyzed to facilitate the process and from NARS and public extension who backstop the technical aspects of the process.

3.4 Access to seed.

Source of seed production always requires particular care and supervision. In traditional local seed systems once the women had selected variety and the planting material, the responsibility for maintaining the true characteristics of a variety, particularly important for household food security and/or socio-economic concern, often rested with the men in seed gardens or specially selected seed plots. This activity is often undertaken by the village seed germplasm repository system, its complexity is determined by socio-cultural factors obtaining in a community. However, changes in socio-political governance of modern communities makes access to technology very difficult, as traditional systems of germplasm transfer have been compromised by modern socio-economic and political factors which have interfaced with the functional cultural organization of farming systems through regulatory framework which deny access to germplasm; for example, compulsory varietal registration and certification.

The National Agricultural Research Systems (NARS) in collaboration with International Agricultural Research Centres (IARCs) now constitute the most viable alternative and sustainable source of both old and new germplasm. IARCs which fall under the auspices of the Consultative Group for International Agricultural Research (CGIAR) have made concerted efforts world-wide to collect, document and conserve the crop bio-diversity of all important ecotypes world-wide. However, the local germplasm repository system usually has no access to such centres, the main constraints being dysfunctional communication systems which prevent effective network linkages between these groups of participants. The constraints identified between farmers and technology and information sources can be divided into the following categories:

- *regulatory*: compulsory variety registration often lock-up valuable material which under commercial farming systems may be legitimate but which are not relevant to local farming systems where heterogeneity in traits and diversity in varieties often underpin the success of a farming system to meet the food and socio-economic demands of the family; and
- *distance*: nearly 70 percent of small scale farmers who rely on local seed systems are to be found in areas with poor and/or inaccessible road and electronic services. This limits interaction between farmers and the service providers.

These constraints have contributed significantly to the decline in crop productivity and increased incidence of food deficits and poverty.

However, The SADC/GTZ project, "Small Scale Seed Production By Self-Help Groups" (SSSP), has successfully launched five pilot informal seed provision systems based on the above concepts in three different SADC countries. Through a network of various projects under its control, SSSP has illustrated that collaborative activities between NARS/IARCs, the farmer, extension services and NGOs can empower farming systems to access germplasm of their preference. While the process requires external stewardship for a year or two, the farmers, through self-help financing, can have access to seed through the network of stakeholders.

4. INTERVENTION STRATEGIES FOR RESTORING PLANT GENETIC RESOURCE

4.1 Participatory variety selection

Participatory variety selection based on technology stored in various gene-banks under the control of institutions affiliated to CGIAR and NARS offer the most cost-effective method for increasing both crop and variety biodiversity. These have a tendency to be restricted in their application by various institutional release processes. A successful programme in the Manica Province was set back three years when it fell foul to restrictive compulsory release processes designed to control entry of hybrid and other exotic material.

It is recommended that the FAO, through NARS, harmonize regulations to permit easy access of IARC-cleared material into the farming systems of countries in the region. There is evidence that several crop varieties have a wide adaptation to environments in neighbouring countries. Testing of such varieties during favourable conditions will create a data base, which can be drawn on regional variety maps which could be used in a Geographic Information system (GIS). With such a powerful tool, it would be possible to intervene effective with appropriate seed of any crop variety before, during and after emergencies either locally or in countries with a similar ecological profile.

There are organisations trying to develop such a system through regional seed networks. Therefore, collaboration of such activities would go a long way in providing a cost effective method of disaster management at international level. It would then be easy to link NGOs and extension services through such a system.

4.2 Strengthening informal seed systems

Programmes designed to empower small scale farmers to be self reliant in seed provision offer the best option for early intervention by philanthropic organisations. Such programmes have been initiated by SSSP in three SADC countries, and linkages with similar in-country programmes appear to be bearing fruit. This suggests, that with slight modifications, a global informal seed provision systems model can be developed. This could be the basis for an international protocol for disaster mitigation, which could then be used in collaboration with GIS.

4.3 Harmonisation of regional seed laws

Seed laws have often been used to restrict access to technology by small scale resource-poor farmers rather than provide proprietary or quality protection. While phytosanitary requirements deserve a lot of respect, restrictions based on genetic purity of open pollinated varieties is often petty.

The conference is urged to recommend that FAO member countries, remove such punitive regulations wherever they exist.

4.4 Institutional capacity building

While most aid organisations have tended to rely on NGOs to implement their projects, the competency of the majority of these organisations, in particular with relation to seed technology, is questionable. The international donor organisations would be best served if an integrated technological delivery system was developed as that which obtains with GTZ-PRRS and DDAP in Sofala Province of Mozambique.

5. REFERENCES

Almekinders, C .J. M., Louwaars, N. P. and de Bruijn, G. H. 1994. Local seed systems and their importance for an improved seed supply in developing countries. In: *Euphytica* 78: 207-216, 1994

Berg, T. 1994. Local Seed Systems Mass Selection How Simple? Plant Breeder/Scientist. NORAGRIC., Agricultural University of Norway. Box 5002 Aas. Norway

Chapman, J., White, J. and Nankam, C. 1997. World Vision's experience with seed supply during emergency and resettlement in Mozambique and Angola: Implications for the future .Pages 147-156. *In:* Alternative strategies for small holder seed supply: proceedings of the International Conference on Options for Strengthening National and Regional Systems in Africa and West Asia..10-14 Mar. 1997, Harare, Zimbabwe. Rohrbach, D. D.. Bishaw, Z, and van Gastel A. J. G. (eds.). International Crops Research Institute for the Semi-Arid Tropics

Eyzaguirree, P. and Iwanaga, M. 1996. (eds.). Farmers' Contribution to Maintaining Genetic Diversity in Crops and its Role Within the Total Genetic Resource System. In: Participatory Breeding. Proceedings of a Workshop on Participatory Breeding, July 1995, Wageningen, The Netherlands. (International Plant Genetic Resource Institute) Rome, Italy.

FAO. 1996. Global Plan of Action. p.16

Ghijsen, H. 1996. The development of varietal testing and breeder's rights in the Netherlands. Pages 223-226. *In:* Integrating seed systems for annual food crops (van Amstel, H., Bottema, J., Sidik, M and van Santen, C., (eds.) Bogor, Indonesia: CGPRT Centre.

Local Seed Systems News. Vol. 3 No. 3 June. 1997. Published by SADC/GTZ Project Promotion of Small Scale Seed Production by Self Help Groups. Harare, Zimbabwe

Longley, K. 1993. Farmers adaptation of rice varieties to different environments in Sierra Leone. *In:* Cultivating Knowledge: Genetic diversity, farmer experimentation and crop research. (Boef, W. de, Amanor, k, Wellard, K., and Bebbington, A. (eds.). London Intermediate Technology Publication

Lourette, D., and Smale, M. 1996. Genetic diversity and maize seed management in a traditional Mexican community: implications for *in situ* conservation of maize. NRG Paper 96-103. Mexico. DF Centro Internacional de Mejoramiento de Maiz y Trigo.

Monyo, E. S. 1998. Analytical Review of Regional Impacts of SMIP: 15 Years of Pearl Millet Improvement in the SADC region. Paper presented at Sorghum and Pearl Millet and Development in Southern Africa: Stakeholders Review and Planning Conference. 27-31 Jul. 1998. Harare. Zimbabwe. SADC/ICRISAT Sorghum and Millet Improvement Programme-Box 776. Bulawayo Zimbabwe.

Musa, T. M., and van der Mheen-Sluijer, J.1977. Review of the state of art in the field of local level seed supply systems. Harare. Published by SADC/GTZ Project Promotion of Small Scale Seed Production by Self Help Groups.

Musa, T. M, and Rusike, J. 1997. *Constraints on Variety Release, Seed Production and Distribution of Sorghum ,Pearl Millet, Groundnut and Pigeonpea*. ICRISAT Southern and Eastern Africa Region Working Paper No 97/02 (SADC/ICRISAT - SMIP). Matopos.

Musa, T. M. 1996. *Pilot Project Sorghum And Pearl Millet Seed Production In Ward 21 Of Chivi, Zimbabwe.* SADC/GTZ Project on Small Scale Seed Production by Self-Help Groups. Harare, 12 pp.

Rohrbach, D. D. 1997. Farmer to farmer seed movements in Zimbabwe: issues arising. Pages 171-179. *In:* Alternative strategies for small holder seed supply: proceedings of the International Conference on Options for Strengthening National and Regional Systems in Africa and West Asia..10-14 Mar. 1997, Harare, Zimbabwe. Rohrbach, D. D.. Bishaw, Z, and van Gastel A. J. G.; (eds.). International Crops Research Institute for the Semi-Arid Tropics.

Salazar, R. 1992. Community plant genetic resources management experiences in southeast Asia. Pages 17-29 *In* Growing diversity: genetic resources and local food security. (Cooper, D., Vellve, R., and Hobbelink, H/eds.) London. Intermediate Technology Publication.

Tripp, R. B. 1977. Between states and markets-innovations for small scale seed provision. Pages 195-210. *In:* Alternative strategies for small holder seed supply: proceedings of the International Conference on Options for Strengthening National and Regional Systems in Africa and West Asia..10-14 Mar. 1997, Harare, Zimbabwe (Rohrbach, D. D.. Bishaw, Z, and van Gastel A. J. G.; eds.). International Crops Research Institute for the Semi-Arid Tropics.

Background Papers

FARMER SEED SYSTEMS AND DISASTER

International Plant Genetic Resources Institute[13]

1. THE SIGNIFICANCE OF FARMER SEED SYSTEMS

It is only in the last decade that farmer seed systems have begun to be recognised as a valuable resource in agricultural development. There exists comparatively little literature systematically describing farmer seed systems and only a handful of studies have closely examined such systems in times of stress. Yet the significance of farmer seed systems to agricultural production cannot be overlooked: farmer seed systems – also referred to as local seed systems or the informal seed sector – provide over 80% of the total amount of seed planted in both developed and developing countries (Cromwell, 1996).

Unlike formal sector seed systems, farmer seed systems are inherently flexible and dynamic: this flexibility often allows them to continue to function in the face of the considerable changes brought about by disastrous events. One of the major findings of ICRISAT's assessment of the emergency response to the 1991/92 drought in southern Africa was that, contrary to assumptions made by the agencies involved in recovery efforts, many small farmers were able to either retain their seed stocks or obtain seed from sources within their communities, despite the extreme severity of the drought (Friis-Hansen & Rohrbach, 1993). Similarly, CIAT's assessment of the impact of war on agricultural production in Rwanda revealed, again contrary to expectations, that only about one-third of farmers had lost any of their local crop varieties and that these farmers could access lost materials through the continued functioning of local seed channels (Sperling, 1996). The apparent resilience of farmer seed systems under difficult conditions suggests that there is a need to understand more fully how such systems operate, both under normal conditions and in times of crisis, and how they can be strengthened to reduce the negative impacts of potential disasters.

2. WHAT DO WE MEAN BY 'FARMER SEED SYSTEM'?

There is a tendency to assume that a farmer seed system refers only to the acquisition of seeds by farmers. Although this forms one of the most important aspects of local seed systems, much of the seed planted by farmers (generally between 50% and 80%) is in fact seed that farmers have kept from the previous harvest. A local seed system thus forms an integral part of the wider agricultural system and depends largely on the capacity of local farmers to plant crops each season and successfully retain some of the output for planting in the following season. In cases where a farmer is unable to retain part of the harvested output, or where a farmer decides to plant a different seed variety, seed is generally acquired from within the local community or within the farmer's wider social network. Another aspect of a local seed system concerns the ways in which farmers manage their seeds as genetic resources: it is possible that particular local agricultural practices may, over many years, lead to changes in the genetic make-up of the seed itself. Such changes have been described as local crop development (Hardon & de Boef, 1993) or farmer breeding (McGuire & Manicad, 1998).

From the above, it can be said that a farmer seed system broadly refers to the processes which farmers use to produce, obtain, maintain, develop and distribute seed resources, both from one growing season to the next and in the long-term. These processes can be examined from social and technical

[13] Prepared by Catherine Longley (Overseas Development Institute) and Paul Richards (Group for Technology & Agrarian Development, Wageningen Agricultural University)

perspectives and in terms of the dynamic ways in which these two perspectives interact. This paper describes aspects of farmer seed systems from the social and socio-technical perspectives.

Three major aspects of a farmer seed system are considered here: (1) the nature of planting material in relation to local agricultural production; (2) mechanisms of seed acquisition; and (3) seed management strategies relating to *in situ* conservation, local crop development and farmer breeding. Each of these aspects is described below, highlighting features of vulnerability and resilience in times of stress and how appropriate assistance might be provided for disaster mitigation and rehabilitation.

Key variables for assessing the impact of disaster on farmer seed systems are presented, and the need for further local-level research is emphasised. The final section contains recommendations concerning the roles and responsibilities of the various agencies involved.

3. ASPECTS OF A FARMER SEED SYSTEM

3.1 Seeds, planting material and local agricultural production

Variations in local agricultural production strategies create considerable contrasts within and between farmer seed systems. The degree of dependence on a single crop (as in Asian wet rice systems) or several major staples (as in swidden systems), together with the biological features of the crops grown, are particularly important technical aspects which may either restrict or allow a range of choices in the ways farmers manage their planting materials. A particular crop's vulnerability to pests, diseases and drought, and its ability to withstand low levels of nutrients or neglect are all biological features that help to determine local cultivation practices and seed systems. Box 1 illustrates how the build up of disease in seed potatoes cultivated in the Peruvian Andes requires that farmers obtain fresh seed every few years.

> Box 1: Potato seed degeneration and regeneration within farmer seed systems in highland Peru
>
> High altitude agriculture brings with it particular problems, for example, risks of drought or of sudden heavy rains which cause flooding and frequent frosts and hailstorms. To reduce risk under such conditions farmers have developed a range of strategies and techniques, important examples of which include the careful management of seed and the maintenance of a wide range of varieties adapted to different conditions and of channels for accessing these varieties. About 90% of the potato production area in Peru is linked in to the informal seed system. This system is characterised by farmers who mainly conserve their own seed from season to season using seed selection and storage techniques. Since virus build-up over the years may cause degeneration of potato seed stocks, particularly in potatoes cultivated at lower altitudes, seed may also be moved between fields at different altitudes as a further means to maintain quality. Farmers periodically obtain fresh seed in small quantities for multiplication in order to eventually replace the older seed of particular varieties. The seed renovation frequency reflects the velocity of degeneration in each agro-ecological zone. Between 3200 and 4000 m (the main production zone) it was found to be 7 years for modern and more for native varieties. Seed for renovation is most commonly obtained from other small farmers living locally or in neighbouring communities. In zones with faster seed degeneration (below 3,000 m) fresh seed is obtained in larger quantities from higher or more isolated zones every two to five years. In these circumstances specific localities can become specialised in informal seed production. In the rough topography of the Andes these interzonal seed flows can occur over quite short distances.
>
> Scheidegger, Prain, Ezeta & Vittorelli, 1989

The length of the crop's growing season and its ratio of seed to consumption are two other important biological factors that influence the local seed system. For crops or crop types with very short growing cycles that can be planted twice or even three times per year, farmers must have access to seed at the appropriate times of year and relatively short periods of seed storage are required. Seed (or other planting material) to consumption ratios vary by crop type, and it is this that helps determine what proportion of the harvest output a farmer must keep for planting in the following season. For example, 30-40% of yam farm output has to be held back as planting material, whereas the equivalent figure for cereals is considerably lower (e.g. rice: c. 5-7%; sorghum: c. 2-3%). For some crop types, such as cassava (manioc), there is no diversion of planting material from consumption: cuttings from the plant itself are planted and the crops reproduce vegetatively.

The crops and crop varieties cultivated by farmers relate to local environmental factors as well as socio-economic considerations such as local livelihood strategies, crop uses and farmers' access to land, labour and capital (including seed). The range of types within a crop - whether plastic (i.e. many adapted types of the same crop) or constrained (i.e. the dependence on a few narrowly diverse varietal types, such as improved, high yielding varieties with heavy chemical dependence) – determine the choice of varieties available to farmers. Within the same crop, farmers have detailed knowledge about the ways in which different varieties display different characteristics, particularly in terms of their different responses to environmental and management features, and there is generally a fairly high rate of turnover in the varieties planted by the same farmer over time as his/her circumstances and needs change.

Changes in local environmental and socio-economic conditions over time lead to changes in agricultural practices and local cropping patterns and consequently both the amount and range of seed types locally produced and/or planted. The dynamism inherent within local agricultural systems, as evidenced by the ways in which farmers alter their practices according to local conditions, is a feature of small scale, semi-subsistence agriculture that becomes especially noticeable when changes occur over relatively short periods of time, as happens in the event of a disaster.

Box 2: Farmers' responses to drought in Bangladesh, 1978-79.

Bangladeshi farmers considered the 1978-79 drought to have been the worst in living memory. It affected three crop seasons, reducing rice production by an estimated two million tons. Farmers responded to the situation by adopting new agricultural practices and by adjusting their crop rotations to compensate for crop losses and reduce the effects of the drought. Where possible, farmers compensated for rice losses by increasing the area under irrigation and by expanding the area under wheat and other dryland crops. There was a considerable expansion of three species of famine millets (probably *Echinochloa colona, E. crisgalli and E. frumentaceum*), especially in western Districts. Other dryland crops such as foxtail millet (*Setaria italica*), finger millet (*Eleusine coracana*) and quick-maturing summer crops, e.g. *Amaranthus* spinaches, were also grown on an increased scale. These innovations and the adjustment of cropping patterns illustrate farmers' flexibility in the face of abnormal conditions.

Brammer, 1987

Changes in agricultural practices and cropping patterns, as illustrated by Box 2, often constitute important ways in which farmers are able to respond to a crisis. Short-season crops, and crops types that ripen earlier, tend to be at a premium in crisis conditions; roots and tubers are also valued by farmers faced by the uncertainties of disasters since they can often be kept in the ground until required for consumption. Agricultural systems with two or three major staples tend to be less vulnerable - especially if the staples come from different groups (cereals, roots, legumes) - than systems heavily dependent on a single staple. Crop substitution is an important feature of multiple staple systems in times of stress. For example, in normal conditions rice in Sierra Leone accounts for 50% of food staple calorie intake and sweet potato and cassava a firther 30%; in war zones the amount of cassava and sweet potato planted tends to rise. It is important for agencies to know about potential areas of substitution, since the first instinct to concentrate on the main crop (rice instead of sweet potatoes in the Sierra Leone case) may not be the only way to solve the problem. Within the same crop, farmers may also substitute high management crop types requiring high labour inputs for low management types requiring less labour.

In disasters where farmers' access to land and labour resources are affected (e.g due to population displacement and /or increased mortality resulting from war, famine or epidemics) the impact on local agricultural production and seed systems can be far-reaching. For example, the rural impact of HIV/AIDS is today a major factor on agriculture in parts of central and eastern Africa, and this impact will intensify through large parts of rural Africa and Asia in coming decades. Rugalema's (1998) study of the impact if HIV/AIDS on rural Bukoba District in North-west Tanzania reveals growing numbers of widow- and child-headed households, abandonment of commercial and labour-intensive crops like coffee and bananas, and much more emphasis on low-intensity cultivation of survival crops like sweet potato and cassava. Such cropping changes are likely to be permanent: unlike war, the HIV/AIDS emergency cannot be ended unless one day a cheap cure is found.

In order to determine local needs and the most appropriate forms of assistance, agencies must be aware of how a disaster alters local agro-ecological and socio-economic aspects of agricultural production and how farmers respond, particularly in terms of new cropping patterns. If seeds or planting materials are to be provided, detailed attention must be given to the types of crops and particular crop varieties required not only in terms of ecological adaptability but also in relation to changing circumstances at local level.

3.2 Mechanisms of seed acquisition

Within local systems, seed is acquired by farmers in three main ways:

(i) by saving seed from the previous harvest;
(ii) as loans, gifts, or other forms of reciprocal assistance (including seed-for-seed exchanges between farmers); and
(iii) by purchasing seed (either for cash or in exchange for other commodities) through markets or localised trade networks.

Research has shown that the quality of seed produced and saved by farmers is maintained at a high level (Wright *et al*, 1995). The importance of seed-saving is such that a change in farmers' capacity to produce and maintain seed is likely to significantly alter the other two main channels of seed acquisition (reciprocity and purchasing). Most systems combine the three modes of seed acquisition in varying proportions depending on crop and varietal types, rates of seed replacement, crop uses, social networks and market integration. Table 1, below, presents data collected from farmers in the Machakos areas of Kenya regarding their seed sources for five crops. The importance of saved seed is clearly revealed.

Table 1: Seed sources as percentage of total seed use in Machakos area, Kenya, short rains 1983.

	Own seed (saved)	Commercial seed (formal sector)	Neighbours	Local market
Maize	83%	12%	1%	4%
Sorghum	77%	8%	8%	5%
Beans	89%	2%	2%	7%
Cowpeas	80%	2%	8%	10%
Pigeon peas	81%	1%	2%	15%

Source: Muhammed *et al*, 1985 in de Bruijn *et al*, 1994.

In contrast, a study of maize seed acquisition among 39 farmers in the valley of Cuzalapa on the Pacific coast of Mexico illustrates the relative importance of seed exchange: for all six cropping seasons surveyed (1989-91), seed from farmers' own harvests represented just 45% of the maize area in the study zone; 55% was planted to seed acquired from other farmers (of which 40% was acquired from within the community and 15% was seed introduced from neighbouring communities) (Louette & Smale, 1996). When these figures are compared to those recorded for maize in Kenya (Table 1), where 83% of maize planted is saved seed, we see that farmer seed systems for the same crop vary greatly according to location.

In the Great Lakes region of central Africa, local markets form an important source of bean seed: surveys conducted in 1991/92 in Zaire, Rwanda and Burundi reveal that about 60% of the farmers surveyed acquired seed from their own production. Various categories of market traders (farmer sellers, small local merchants and large merchants) formed the other significant seed source, as illustrated in Table 2, below. It was found that poorer farmers rely more heavily on seed from the market than richer farmers, who tend to use the market only as a way of accessing specific bean varieties.

Table 2: Great Lakes region: percent of farmers who relied on a particular bean seed source during the principal growing season, 1991-92. (Figures exceed 100% as farmers may use multiple sources.)

	Own stock (saved)	Relatives & neighbours	Local market sources
Zaire (n=194 farmers)	59%	1%	59%
Rwanda (n=144 farmers)	63%	10%	32%
Burundi (n=248 farmers)	66%	5%	50%

Adapted from Sperling, Scheidegger & Buruchara, 1995.

The categories defined in Box 3, below, reveal how socio-economic differentiation among individual farmers and farming households influence the ways in which farmers within the same community access seed. The different livelihood strategies and household responsibilities of different farmers (e.g. according to age and gender) may also partly determine how seed is acquired. For example, variations in seed sources for male and female rice cultivators from two neighbouring ethnic groups in northern Sierra Leone, as presented in Table 3, reflect Susu women's greater involvement in trade activities. For both ethnic groups, women tend to rely more than men on acquiring seed through loans, gifts and reciprocal assistance.

Table 3: Sources of swamp rice seed planted in two neighbouring villages, northern Sierra Leone, 1994. Figures refer to seed amounts as proportions of total seed planted. n=53 farmers.

	Farmer-saved seed	Seed acquired through loans, gifts or reciprocity	Seed purchased from local traders
Susu women	43.8%	37.5%	18.8%
Susu men	75.3%	24.8%	0
Limba women	67.2%	32.8%	0
Limba men	67.5%	19.7%	12.8%

Longley, 1998b

Box 3: Household categories according to principle modes of seed acquisition

Individual households and farmers use different sources of seed over time, resulting in complex patterns of seed sourcing. Broadly-speaking, at any given point in time three different categories of households can be distinguished:

Seed secure households (usually the better resourced households), who save the majority of seed needed on-farm but may use off-farm sources (from relatives, neighbours and markets) in order to experiment with new varieties;
Crisis-prone households who are seed-secure in most seasons but are less well-resourced, so a domestic crisis, such as harvest failure or death or illness in the family, can force them to augment saved seed with off-farm supplies;
Chronically insecure households who are poorly resourced and so in most seasons are unable to harvest sufficient crops to meet seed and domestic consumption needs.

Cromwell, 1996.

Farmers' seed sources change over time in response to local needs and circumstances, as illustrated by Box 4. When disaster strikes, it is often those who are least able to access seed in normal times who suffer the most in terms of reduced seed access in times of crisis, as illustrated by the case of refugee farmers from northern Sierra Leone (see Box 5). Such findings suggest that the targetting of seed and other forms of emergency aid according to socio-economic differentiation and farmers' ability to access seed may be desirable were sufficiently detailed data on emergency needs can be compiled. Due to the difficulties experienced by chronically insecure households in saving seed from one season to the next, seed assistance targetted to such farmers may have less of a long-term impact than that targetted to those better able to save seed (Longley, 1998a). Although such farmers require seed in the short term, other forms of aid – particularly those that address more general poverty needs – may be more appropriate for the long term.

Box 4: War, drought and farmer seed systems in Zimbabwe

As Zimbabwe's liberation war escalated in the 1970's, rural farming communities were forced to live in so-called "protected villages" to restrict their movements and to reduce ground support for the guerilla fighters. Black farmers who lived through this period recount that great loss of local varieties occurred at this time since they were under curfew and could not guard their crops against pests and raids. In addition, they had been unable to move their granaries and seedstocks during the relocation. The restriction on farmers' movements also disrupted local patterns of seed distribution and when seed stocks of particular traditional varieties were depleted these could not easily be replenished from mothers and other relatives, as usually occurs. However, it is likely that seed from neighbouring countries particularly Mocambique and Zambia, which provided safe haven for Zimbabwean guerillas during the war, were transported with the guerillas and were grown in Zimbabwean villages.

When independence was achieved in 1980, the new government immediately sought to increase agricultural production among black farmers. A number of agricultural research stations were opened in the communal areas and crop packs, containing hybrid maize seeds and fertiliser, were handed out to all non-commercial farmers. A decade later, in the throes of the century's worst drought, farmers lamented the loss of their traditional drought-tolerant varieties and ascribed this loss to the heavy promotion of hybrid maize and free crops packs which was begun at independence.

Van Oosterhout, 1996

Box 5: Farmers' responses to war and insecurity in northern Sierra Leone, 1994-97.

Data collected both before and after rebel attacks and population displacement among almost 150 Susu farmers living near the Guinea border in northern Sierra Leone show that the 49 sample farmers (20 women and 29 men) who transferred to refugee camps over the border experienced considerable difficulties both in producing and saving suitable seed for planting. Refugee farmers reported a direct causal link between problems of access to fertile land and their ability to produce seed: their crop outputs were lowered due to both the small size of the land available to them and the infertility of the soils. Lowered outputs are more difficult to manage – not only is there not enough grain to feed the family but due to food shortage it becomes very difficult to avoid eating what would normally be kept as seed. Yet many of these 49 farmers were from chronically insecure households who, even before the crisis, planted consistently lower proportions of self-saved seed than the other farmers surveyed. Those farmers with least access to self-saved seed in normal times faced even greater problems brought by the crisis and were forced to move to the refugee camps. Although the proportion of seed saved by farmers who remained in their original homes also declined as a result of theft and disruptions to farming activities, these farmers were better able to maintain seed from one year to the next, both before and during the crisis period.

Longley, 1998a.

If assistance to support local seed systems is to achieve longer-lasting impacts, it is necessary for agencies to pay greater attention to the social aspects of such systems. The vulnerability and resilience of local seed systems need to be seen not just in terms of the physical survival of seeds and storage, but in terms of how the social basis for seed acquisition is affected by emergency conditions. For example, farmers and merchant traders impoverished by war or drought may lack money to re-enter markets; civil strife and inter-communal violence, as in Rwanda or Bosnia, may destroy the basis of trust between neighbours necessary for loans or forms of reciprocal assistance to function (cf. Richards 1996;1998); in cases where farmers are displaced or move into refugee camps, previously efficient social networks may become physically scattered. HIV/AIDS may lead to the social exclusion of afflicted individuals and family survivors and seed systems may be especially badly disrupted where key commercial or technical functionaries (e.g. seed merchants, rural transporters, agents for government input supply chains) are affected by the disease, particularly since asset-disposal by those with HIV/AIDS is often very dramatic (Rugalema 1998). Unlike drought, but more like war, HIV/AIDS does not differentiate between marginals and larger farmers at the heart of local seed acquisition mechanisms.

The role of village grain traders as sources of credit in kind, and of older more established farmers (women as well as men) as agents of informal inter-seasonal seed exchange is often vital to the proper functioning of a village seed system. Like HIV/AIDS, war targets the human capital upon which such informal exchange arrangements depend. The rebel Revolutionary United Front in Sierra Leone endeavoured to gain popularity by killing or driving out village produce traders; a move it thought would "wipe out" farmers' indebtedness, but without any understanding of the consequences for local seed systems. The effect is not unlike the results obtained from NGO-led projects in drought-affected areas of the Sahel to replace village traders with "community seed banks". These frequently prove unsustainable, but traders have been meanwhile undermined or put out of business. Sellies and Wanders (1997) arrive at the rather depressing conclusion that whereas trader's margins are often high, farmers lose even more when they rely upon friends and kin to sell their produce. Renovating a local seed system may depend on bringing the traders back in, and seeking to boost their efficiency through increased competition. The key to this might be rehabilitation of rural roads, and recapitalising locally-based seed merchants through rural bank loans (and providing the softest loans to traders prepared to re-settle or continue to work in more isolated, less secure or more HIV/AIDS afflicted communities), rather than direct intervention in seed provision.

Associated with appropriate support for community seed providers should be support to less well-resourced households and individuals so that they are better able to access seed available within the community. Timely access to the right kinds of planting material often depends upon having relevant material to exchange for seed at the right moment. As pointed out in Amartya Sen's work on exchange entitlements in famines (1981), some crises occur more through the loss of access to material with which to acquire the right inputs than through direct shortage of the inputs. This was one of the major findings of the SOH assessment, which suggests that poverty-based approaches may provide greater opportunities for poorer farmers to access seed (Sperling, 1996). It may be as important to concentrate on supplies of exchangeable items as on seed itself.

3.3 Seed management, in situ conservation and local crop development

It is now recognised that the maintenance of genetic diversity is essential for local farming systems and the long-term sustainability of agricultural production in complex and changing environments (FAO, 1996; ODI, 1998). There is a rapidly growing body of literature that examines how farmers' crop and seed management strategies influence the local conservation, erosion and evolutionary development of crop genetic resources (e.g. deBoef et al, 1993; van der Heide et al, 1996; Sperling & Loevinsohn, 1996). As mentioned in Section 3.1, farmers often maintain a range of crops and crop varieties as a way of coping with changing circumstances. The introduction of new varieties from outside the community depend upon the ability of individuals to travel outside their home settlement to acquire seed from relatives, markets and other sources, as illustrated in Box 4. The acquisition of new varieties also depends on the proper functioning of traditional forms of gift-giving, the availability (and affordability) of transport services and an economic infrastructure to support the marketing of goods and services. The maintenance of seeds and genetic variability within a farming system are closely linked with the particular features of local social organisation, economic activities, and cultural/religious norms of the community, in addition to agro-ecological considerations (Brush, 1986; Dennis, 1987; Longley, 1998; Soleri & Cleveland, 1993; Zimmerer, 1992).

Whilst disasters are generally thought to have a negative impact on genetic diversity, there are a handful of cases to suggest that the changes brought about by a crisis may also promote local diversity. For example, in local agricultural systems where there is varietal plasticity, varietal turnover rates tend to be quite high. This process of change may be further increased in times of crisis. In Rwanda, farmers had been growing beans in mixtures long before the events of 1994. Yet it would seem that the war and its associated effects resulted in the intensification of varietal mixing, creating unusually dynamic varietal profiles (Sperling, 1997). Farmers' responses to disasters may promote changes in local diversity by evolutionary processes, as illustrated by Box 6. Although considerably more research is needed to fully understand the impact of disasters on local crop diversity, the apparent increase in the variability of crops cultivated during times of stress may support the need for emergency germplasm collection activities in times of crisis.

Box 6: Drought and potato evolution in the Andes

A study into the biological and cultural factors determining the ongoing evolution of the potato in the Andes of western Bolivia show that humans have directed evolution on small scale and that human intervention in evolutionary processes may be more deliberate during time of crisis. The formation of hybrids between various potato species is an important phenomenon in the Andes and ongoing introgressive hybridization is most likely to occur in places where abundant wild and domesticated plants grow in close association. A valley known as Ajawari Marka (lit. 'the place where the domesticated wild potato grows') adjacent to the town of Caquiqviri in the Department of La Paz, Bolivia, represents such a place. In western Bolivia, periodic droughts and crop failure necessitate the replenishment of potato stocks. The drought of 1982-83 represented the most severe example of such a phenomenon in this century: harvests in the Caquiaviri area were approximately 25% of the normal and it was not uncommon for some farmers to have harvested nothing at all. While the rest of the area was ravaged by drought and frost, yields in the protected valley of Ajawiri Marka, while significantly reduced, were greater than anywhere nearby. When supplies of the cultivated potatoes were depleted, weed potatoes would be the obvious sources of tubers for food and seed. Under severe ecological conditions in the past, Ajawiri Marka likely represented a reservoir of seed potato from which important local cultivars originated.

Johns & Keen, 1986.

4. KEY VARIABLES IN ASSESSING THE IMPACT OF DISASTER ON FARMER SEED SYSTEMS

The design of appropriate interventions for disaster mitigation and recovery requires detailed information of local level constraints and opportunities. A seed system under stress is not so much a complete transformation of the 'normal' seed system as a system in which certain 'normal' activities have become restricted and others intensified. It is therefore important to have some understanding of local seed management strategies under 'normal' conditions, and particularly how the seed system is shaped by different livelihood strategies and other local socio-economic charateristics of the population. The adaptive responses of an individual or a household to a crisis are determined not only by the nature of the crisis itself but by the particular needs and capacities of those affected and the functioning of the social networks of which they form a part.

Whilst there is an urgent need for further research in order to understand more fully how disasters affect local seed systems, three groups of variables may be useful in assessing farmer seed systems in times of stress: (i) features of the crisis; (ii) the responses of farmers; and (iii) social and economic changes. Various observations relating to each, particularly (ii) and (iii) have already been mentioned above. It is especially important to note that there is often a high degree of local specificity for each variable described: the effects of disaster are rarely uniform across different locations.

4.1 Features of the crisis

Different types of disaster affect farming communities in different ways. Drought, for example, tends to impact hardest on the most marginal households within a community (e.g. landless labourers, migrants with weak entitlements to local social security including seed exchange, part-time cultivators from pastoral groups): an implication is that the means of seed acquisition will not necessarily be disrupted, provided the local population is not driven to seek aid in camps. War generally has more serious consequences for seed systems than drought in that it is either indiscriminate in selecting victims, or actually targets the merchants and wealthy farmers who are often central in maintenance of local seed trade and reciprocal exchange, as described above.

The type, timing, duration and intensity of a crisis are features that influence the response of farmers and types of social and economic change. The timing and duration of a crisis, particularly those with a rapid onset, can be crucial in terms of the level of disruption to agriculture. In war situations, the broad scale of crop and seed losses can be assessed in terms of the actual time that households and farmers are prevented from farming and at what periods of the agricultural calendar (i.e. planting, weeding, harvesting) (Longley, 1998a; Pottier with Wilding, 1994).

4.2 The responses of farmers

Depending on the nature of the crisis and its effects, farmers may: (a) move from their homes to camps for refugees or displaced people; (b) move from their homes to stay with friends and relatives elsewhere; or (c) remain in their homes. In some cases, farmers may move more than once. Which farmers move? Which farmers remain? Who goes where, and when? Answers to such questions relate to the location of the community in relation to areas of risk/safety; regional social networks and potential places of refuge; livelihood opportunities at home/with friends/in camps; inter-household differentiation: access to resources, relative dependence on cash income/agricultural subsistence; and intra-household composition (e.g. gender/age). Factors such as these help to understand the socio-economic conditions of farmers in each location type (see Box 5). In cases where farmers move away from their homes, the timing of their displacement in relation to the agricultural cycle and farm tasks, the duration of time spent away from home and the timing of their return in terms of the agricultural year are factors to be considered in the level of disruption to local agricultural systems and seed production. The distance over which people move, the differences in the agro-ecology and the presence or absence of established social networks in the new location affect farmers' ability to access seed and take up farming elsewhere.

The responses of farmers to a crisis situation also involve changes in their agricultural practices and cropping patterns, as described in Section 3.1. Saved seed and other inputs, particularly labour, may become a constraint to local farming practices, and farmers may compensate for this by acquiring seed from other local sources or by altering their crops and crop varieties to types that require less labour. Different crops are affected by disaster in different ways (Sperling, 1997) and farmers may alter their cropping patterns by substitution of crops and crop types. Quick-maturing types and roots and tubers often become especially important.

4.3 Social and economic changes

Seed acquisition systems are social systems and their vulnerability and resilience needs to be seen in terms of how the social basis for seed exchange is affected by social and economic changes effected in emergency conditions. These changes may include those listed below.

4.3.1 Farmers' **access to agricultural inputs**, particularly land, labour and capital may be reduced: where farmers are displaced over some distance they can no longer cultivate their own land, either temporarily or permanently; the size of land they previously cultivated may be significantly reduced; displacement scatters families and farm labourers; increased poverty may prevent farmers from hiring labourers. Factors such as these affect agricultural production and the types of seeds required.

4.3.2 Poverty may be increased by adverse impacts on **trading and off-farm incomes:** in turn farmers may not have cash or other commodities required for seed acquisition. The importance of farming and subsistence production in relation to other economic sectors (e.g. wage labour or commercial agriculture) may increase. In other situations, opportunities for trade and off-farm earnings may increase for certain sectors of the population.

4.3.3 Changes in the patterns of **social cooperation** among farmers through which seed systems are managed are particularly evident in war situations where there is a loss of trust within and between local communities (Richards & Ruivenkamp, 1997).

4.3.4 The **local mobility** of farmers may be reduced in areas where it is not safe to travel or where transport systems have been affected, affecting access to seed from outside the local area and preventing the movement of labour. In Rwanda, it has been argued that the disruption to the flow of seasonal casual labour has been at least as serious as the damage inflicted by the loss of seeds and tools (Pottier with Wilding, 1994). Farmers may not be able to transport their commercial crops to market.

4.3.5 The **population composition** (according to gender and age) of local communities – whether in refugee camps or in their original homes – may be significantly altered. Men, women and children often have specialised roles in agricultural production and seed systems (in specific labour tasks, selecting or storing material, in knowledge of different crops), and crisis conditions often require that age- and gender-specific skills come together in different ways, e.g. by throwing emphasis on relatively neglected "women's crops", or crops that are mainly used for cash. Careful gender-sensitive and age-sensitive analysis is needed in establishing policy frameworks for seed system rehabilitation.

5. STRENGTHENING FARMER SEED SYSTEMS FOR IMPROVED DISASTER PREPAREDNESS AND DISASTER PREVENTION

The localised nature of farmer seed systems makes it very difficult to make generalised recommendations for strengthening such systems. There is a need for rural development agencies, agricultural research institutes and others to develop better understandings of how local seed systems operate under normal conditions and how they react to pressure. Agricultural development strategies and the national seed sector should seek to enhance rather than undermine farmer seed systems. Crop researchers must give greater priority to the analysis of genetic materials flowing in farmer seed systems (e.g. whether farmer gene pools are rich or poor, whether HYVs are enriched by geneflow from local material, whether HYVs enrich local gene pools, etc) in order to better assess the impacts of disaster on local crop diversity (e.g. whether there is genetic erosion and/or increased variability). Disaster-prone areas, where farmers have well-developed local strategies for dealing with crisis situations, offer enormous potential for developing a better understanding of the capacities and vulnerabilities of farmer seed systems in times of stress (e.g. van Oosterhout, 1996). Regional and national organisations in such areas should promote further research of this kind.

6. INSTITUTIONAL ROLES AND RESPONSIBILITIES

6.1 There is a need for **gene banks** to collect, screen, document and conserve farmer landraces suited to cultivation in difficult conditions, material threatened with loss during crises, and types required for post-crisis recovery. Collection programmes should acquire material with characteristics suited to the restocking of local seed systems in emergencies, and passport data for accessions need to be extended to include relevant information about social uses (including gender significance), and practical utility of material in difficult conditions and emergency situations. Regional and international networking activities must be developed among gene banks to ensure the establishment of and access to back-up collections of farmer varieties of main food security crops.

6.2 Public-interest **agricultural research institutes** (IARCs and NARS) should place greater emphasis on developing seed materials suited to cultivation in difficult conditions, including new types adapted to drought, or to the needs of farmers in disaster areas. The current interest in participatory selection and other decentralised, client-oriented improvement methods may be useful in promoting such research and improvement programmes (see Eyzaguirre & Iwanaga, 1996; McGuire & Manicad, 1998). On-going work at Rokupr Rice Research Station in Sierra Leone with assistance from the Community Biodiversity Development and Conservation (CBDC) Programme and WARDA on the collection, screening and selection of early and hardy African rices and African/Asian hybrids especially suited to farmer resettlement in zones of post-war recovery is an example of such a programme (Jusu, 1999).

6.3 In providing emergency seed inputs to farmers, **NGOs and other relief and development agencies**, should pay greater attention to the technical, organisational and social issues involved. The experiences of a number of NGOs with particular expertise, together with available guidelines (e.g. ODI, 1996; DEFSIL, 1996), can provide useful advice in this area. More emphasis is required in sustainable capacity-building activities to support the post-disaster rehabilitation of farmer seed systems. Greater sensitivity to farmer seed systems is required in the provision of emergency food aid to prevent farmers from having to eat their local seed stocks.

6.4 There is the potential for **farmers and refugee groups** to become directly involved in seed recovery, screening and selection activities with gene banks and research; and to work with NGOs in organising collection, multiplication and dissemination of planting materials for seed system rehabilitation. Where refugee camps are safe and stable much more work could be done at an early stage to involve refugees and displaced farmers in documenting how seed systems work and planning early stages of eventual rehabilitation (including camp-based seed selection and multiplication work).

6.5 A key issue is **institutional collaboration**: how to get the various players listed above to work closer together on seed system security issues. A programme of networking activities, workshops, and joint practical projects needs to be planned at national, regional and international levels explicitly to encourage the emergence of novel forms of collaboration. National agencies that are in a position to promote and encourage such collaboration include agricultural ministries, seed services, and national disaster preparedness committees. The new tendency for humanitarian agencies to pay attention to seed system and plant population genetic issues is welcome, but needs to be further encouraged. The same could be said for moves within public-interest agricultural research to address food security through decentralization, and to place greater emphasis on gender, difficult and marginal environments, and better performance of crop types under drought and other forms of stress. Yet much remains to be done at the level of NARS, other national agencies and agriculturally-oriented NGOs working with low-resource farmers. The gene banking capacity of NARS remains a bottleneck issue: gene banking is the defence of final resort in disasters and needs to be planned on regional and international network lines (Richards & Ruivenkamp, 1998).

7. REFERENCES

de Boef, W., K. Amanor, K. Wellard with A. Bebbington, 1993. *Cultivating knowledge: genetic diversity, farmer experimentation and crop research*. London: IT Publications.

Brammer, H. 1987. Drought in Bangladesh: lessons for planners and administrators. *Disasters* **11** (1): 21-29.

De Bruijn, G., C. Almekinders & L. Keune, 1994. Farmers' knowledge and practices in smallholder seed production, with special reference to a case study in Central America. Paper presented at the ILCA/ICARDA research planning workshop on smallholder seed production, 13-15 June, 1994. Addis Ababa.

Brush, S. 1986. Genetic diversity and conservation in traditional farming systems. *Journal of Ethnobiology* 6(1): 151-167.

DEFSIL Project of USAID/G/ENV/ENR, 1996. Seeds for disaster mitigation and recovery in the Greater Horn of Africa. Report prepared by Chemonics International & USDA Famine Mitigation Activity; submitted to USAID/BHR/OFDA/PMPP.

Cromwell, E. 1996. *Governments, seeds and farmers in a changing Africa*. Wallingford, UK: CAB International, in association with ODI.

Dennis, J. V. 1987. *Farmer management of rice variety diversity in northern Thailand*. PhD Thesis, Cornell University.

Eyzaguirre, P. & M. Iwanaga (eds.) 1996. *Participatory plant breeding*. Proceedings of a workshop on participatory plant breeding, 26-29 July, 1995, Wageningen, The Netherlands. Rome: IPGRI.

FAO, 1996. *Report on the state of the world's plant genetic resources for food and agriculture*. Prepared for the International Technical Conference on Plant Genetic Resources, Leipzig, Germany, 17-23 June, 1996. Rome: FAO.

Friis-Hansen, E. & D. Rohrbach, 1993. SADC/ICRISAT 1992 drought relief emergency production of sorghum and pearl millet seed: impact assessment. ICRISAT Working Paper 93/01.

Hardon, J. & W. deBoef, 1993. Linking farmers and plant breeders in local crop development. In W. deBoef *et al* (eds.). *Cultivating knowledge: genetic diversity, farmer experimentation and crop research*. London: IT Publications.

van der Heide, W., R. Tripp with W. de Boef, compilers. 1996. *Local crop development: an annotated bibliography*. IPGRI, Rome / CPRO-DLO (CGN), Wageningen / ODI, London.

John, T. & S. Keen, 1986. Ongoing evolution of the potato on the Altiplano of western Bolivia. *Economic Botany*, **40**(4): 409-424.

Jusu, M. 1999 (in preparation). *The genus Oryza: sources and uses of genetic variability by Susu and Limba farmers in Sierra Leone*. PhD Thesis, Wageningen Agricultural University.

Longley, C. 1997. Effects of war and displacement on local seed systems in Sierra Leone. In L. Sperling (ed.) *AgREN Network Paper* No. **75**. ODI.

Longley, 1998a. Seed management strategies of farmers affected by war: a case study from northern Sierra Leone. Report to ESCOR, UK Department for International Development.

Longley, 1998b. *A social life of seeds: local management of crop variability in north-western Sierra Leone*. PhD Thesis, University of London.

Louette, D. & M. Smale, 1996. Genetic diversity and maize seed management in a traditional Mexican community: implications for in situ conservation of maize. *CIMMYT Paper* 96-03.

McGuire, S. & Manicad, G. 1998. Technical and institutional issues in participatory plant breeding from the perspective of farmer plant breeding: a global analysis of issues and current experience. *SWP/PRGA Working Paper*. Cali: CIAT.

Muhammed, L., F. Scott & M. Steeghs, 1985. Seed availability, distribution and use in Machakos district, short rains 1983 – short rains 1984. MALD: DAO and NDFRS, Machakos, Kenya.

Overseas Development Institute (ODI) Seeds and Biodiversity Programme, 1996. Seed provision during and after emergencies. Relief and Rehabilitation Network; *Good Practice Review 4.* London: ODI.

ODI, 1998. Agricultural biodiversity and poverty reduction: issues and entry points. Draft report for UK Dept for International Development project: Linking Policy with Practice in Biodiversity.

van Oosterhout, S. 1996. What does in situ conservation mean in the life of a small-scale Zimbabwean farmer? Examples from Zimbabwe's communal areas. In L. Sperling & M. Loevinsohn (eds.), *Using diversity: enhancing and maintaining genetic resources on-farm.* New Delhi: IDRC.

Pottier, J & J. Wilding, 1994. Food security and agricultural rehabiliation in post-war Rwanda, August – September 1994. Report to Save the Children (SCF) UK, October, 1994.

Richards, P. 1986. *Coping with hunger: hazard and experiment in a West African rice-farming system.* London: Allen & Unwin.

Richards, P. 1996. *Fighting for the rain forest: war, youth and resources in Sierra Leone*. The International African Institute in association with James Currey, Oxford & Heinemann, Portsmouth.

Richards, P. & G. Ruivenkamp, with contributions from R. van der Drift, M. Gonowolo, M. Jusu, C. Longley & S. McGuire, 1997. *Seeds and Survival: Crop genetic resources in war and reconstruction in Africa*. Rome: International Plant Genetic Resources Institute (IPGRI).

Rugalema, G. H. R. 1998 (forthcoming). *When disaster unfolds: coping with HIV/AIDS in smallholder agriculture in Bukoba District, Tanzania.* PhD Thesis, Wageningen Agricultural University.

Scheidegger, U., G. Prain, F. Ezeta & C. Vittorelli, 1989. Linking formal R & D to indigenous systems: a user-oriented potato seed programme for Peru. *AgREN Network Paper* **10**. London: ODI.

Sellies, F. & Wanders, J. 1997. *Who pays the price?: the production and marketiung of coffee and cocoa in Sierra leone under structural adjustment.* Saarbruecken: Verlag fuer Entwicklungspolitik.

Sen, A. 1981. *Poverty and famines: an essay on entitlements*. Oxford: Oxford University Press.
Soleri, D. & D. Cleveland, 1993. Hopi crop diversity and change. *Journal of Ethnobiology* **13**(2): 203-231.

Sperling, L. 1996. Executive summary and reflections of Seeds of Hope socio-economic analyses in Rwanda: the impact of war on agricultural production. CIAT report presented to IDRC.

Sperling, L. (ed.) 1997. War and crop diversity. *AgREN Network Paper* No. **75**. ODI.

Sperling, L. & M. Loevinsohn (eds.) 1996. *Using diversity: enhancing and maintaining genetic resources on-farm.* Proceedings of a workshop held on 19-21 June, 1995, New Delhi, India. New Delhi: IDRC.

Sperling, L., U. Scheidegger & R. Burachara, 1996. Designing seed systems with small farmers: principles derived from bean research in the Great Lakes region of Africa. *AgREN Network Paper* No. 60. London: ODI.

Wright, M., L. Delimini, J. Luhanga, C. Mushi & H. Tsini, 1995. The quality of farmer saved seed in Ghana, Malawi and Tanzania. Project A0266. Chatham: Natural Resources Insititute.

Zimmerer, K. 1992. The loss and maintenance of native crops in mountain agriculture. *GeoJournal* 271: 61-72.

Background Papers

PLANT GENETIC RESOURCES AND SEED RELIEF

Toby Hodgkin, Murthi Anishetty
IPGRI,Rome, Italy; FAO, Rome, Italy

1. INTRODUCTION

In the modern world, especially in developing countries, people are threatened with, and vulnerable to, natural disasters, civil strife and war. Such calamities present huge challenges to the resilience of agricultural systems where planting, maintenance and harvesting may all be disrupted or prevented altogether. Often, adapted crop varieties are lost and cannot be recuperated locally. This is particularly damaging for the 1.4 billion people who are estimated to live in farm families that are still largely self-provisioning in terms of their planting materials or seeds.

In 1996, over 150 countries adopted the FAO Global Plan of Action for the Conservation and Sustainable Utilization of Plant Genetic Resources in Food and Agriculture (FAO, 1996a). The main aims of the Global Plan are:

- to ensure conservation of plant genetic resources for food and agriculture as a basis for food security;
- to promote sustainable utilization of plant genetic resources, in order to foster development and reduce poverty and hunger;
- to promote a fair and equitable sharing of benefits from the use of plant genetic resources;
- to assist countries and institutions responsible for conserving and using plant genetic resources to identify priorities for action; and,
- to strengthen plant genetic resources programmes and enhance institutional capacity.

The Global Plan recognized the potential value of the world's plant genetic resources in assisting farmers and communities confronting disasters and identified assisting farmers in disaster situations to restore agricultural systems through the use of plant genetic resources as one of its 20 key activities. The long term objective of this activity was described as supporting "farmers' and rural peoples' livelihoods and sustainable agriculture options through the rehabilitation of agricultural systems based on locally adapted genetic resources, including the restoration of pre-existing germplasm in cases of disaster induced loss of plant genetic resources".

Plant genetic resources provide the biological basis of food security. They include commercial plant varieties (both currently available and those developed in the past) as well as traditional local plant varieties that continue to be an integral part of many production systems throughout the developing world. They also include wild and weedy relatives of useful plant species, many of which have provided valuable characters such as disease and pest resistance which are essential elements of modern varieties.

Based on the work of early pioneers such as N.I. Vavilov and Harry Harlan, crop geneticists and plant breeders of the 1950s and 1960s, who were becoming increasingly concerned at the continuing loss of traditional crop varieties and their wild relatives, began a series of initiatives to support the exchange and maintenance of plant genetic resources. A series of meetings and initiatives sponsored by FAO paved the way for the development of a global programme of support for the conservation and use of plant genetic resources and for the development of national programmes in very many countries throughout the world. As a result of this it is now estimated that there are over 1,400 collections of plant genetic resources maintaining over 6 million accessions (an accession is plant sample, strain or population held in a genebank). The collections occur throughout the world and cover a very wide range of crops including cereals, legumes, roots and tubers, vegetables, fruits and industrial crops.

Maintenance of plant genetic resources is undertaken by a range of different national institutions, which include dedicated genetic resources institutes and other national agricultural research centres, and by several international and regional research centres. FAO provides an international framework for addressing key issues of conservation and use of plant genetic resources both through its regular programme and through the Commission on Genetic Resources. The System Wide Genetic Resources Programme (SGRP) of the Consultative Group for International Agricultural Research (CGIAR) coordinates work by international agricultural research institutes. One institute of the CGIAR - the International Plant genetic Resources Institute (IPGRI) - is wholly concerned with conservation and use of plant genetic resources (FAO, 1996b).

The activity described in the Global Plan took account of a growing concern to ensure that seed provided following a disaster genuinely meets the needs of the recipients. This concern has also been reflected in a number of actions in different parts of the world where the needs of disaster affected peoples for specific types of seeds have been deliberately addressed. The Seeds of Hope programme in Rwanda, which involved a collaboration between International Agricultural Research Centres (IARCs), NARS and NGOs, monitored varietal diversity in Rwanda with the aim of reintroducing lost local varieties (Sperling, 1995). Studies of ways of meeting varietal needs and the provision of desired local types have been undertaken in W. African rice producing areas affected by war (Richards and Ruivenkamp, 1997), In Somalia a collaboration involving local communities, an NGO and IPGRI sought to make lost local types available to farmers (IPGRI, 1998).

There are two dimensions to the provision of varieties to farmers following some type of disaster. One concerns the identification of needs based on an understanding of production systems and the effect of the disaster, particularly in respect of the material needs of the producers. The other concerns the capacity to meet those needs through the provision of the most suitable planting materials. The first aspect necessarily involves some estimation of what needs to be restored or replaced (demand) while the second aspect is concerned with the material and capacities that exist at local, national or international levels that can be mobilized to meet the identified needs (supply). This paper is largely concerned with the latter aspects - specifically with the plant genetic resources that are available as a result of the global conservation efforts of the last 50 years, how one might be able to identify and use them following disasters, and how their contribution might be maximized. The objective is to explore both the contribution that the materials themselves may make and to consider the ways in which the methods and approaches developed for conservation of plant genetic resources may be useful.

In the first part of this paper we draw on the FAO Report on the State of the Worlds Plant Genetic Resources (FAO, 1996b) to describe the materials that are conserved as plant genetic resources for food and agriculture, their characteristics, the procedures that are used for their maintenance and the types of information that may be available on the resources conserved. The types of institutions that are involved in maintaining plant genetic resources are also described. In the second part of the paper we explore some of the factors involved in making these resources available for relief following various types of disasters. The concern is primarily with technical issues involved in identifying the most useful sources of information, the institutions that may be able to provide materials of relevant information and the ways in which useful materials could be located. In the final section we identify some issues involved in developing improved systems and capacities.

2. THE CONSERVED RESOURCES

Over the last 40 years there has been a substantial replacement of the traditional varieties of the worlds' major staple crops (e.g. wheat, rice, corn, sorghum and major pulses) with new improved varieties produced as part of national or international publicly or privately funded plant breeding programmes. It is now estimated that over 80% of the area of developing world bread wheat production is sown to modern varieties (CIMMYT, 1996). However, this global figure masks substantial regional and national or local differences; while the area sown to modern varieties in South Asia is over 90% that in West Asia is only 53%. The situation is similar for rice, with 74% of the total Asian area planted to modern varieties, although in Cambodia only 11% of the area is sown to modern varieties, while in Malaysia the coverage is 90%. In contrast to rice and bread wheat, modern varieties still account for less than 50% of the production of pearl millet, even in countries such as India where substantial breeding is in progress. In some African countries the reliance on traditional varieties remains very high for a number of staple crops such as millet, sorghum and rice.

A major concern of plant genetic resources conservation programmes is to ensure that the genetic diversity present in local varieties is maintained for use in future breeding programmes. It has always been recognized that they contain many valuable characters (e.g. disease and pest resistance, stress tolerance) and constitute a vital resource for future crop improvement work. The emphasis of early work was on collecting and maintaining these materials and in establishing national or regional genebanks concerned with their maintenance and distribution. As a result of this, there are now very substantial collections of many crop species. Procedures for their optimum maintenance over long periods have been developed together with information management systems to keep track of the materials and of their attributes. Distribution problems have been tackled and countries have developed a range of approaches to ensure that work within a country is adequately managed and makes the best possible contribution to national needs.

Most of the work of national programmes on conservation of crop plant genetic resources is concerned with *ex situ* conservation. However, while *ex situ* conservation provides an essential national and global resource, much diversity continues to be found in the production system itself. The earlier view that new improved varieties will always rapidly replaced traditional ones is now seen to be oversimplified and often unfounded. There appear to be many situations and circumstances where farmers continue to maintain and use traditional varieties because these best meet their current needs. From the conservation perspective, it is therefore clear that the maintenance of diversity in production systems (conservation through use) should continue to be a key element in the global conservation effort together with deliberate and planned actions to conserve genetic resources *ex situ* (conservation for use). In the context of response to disaster situations, both *ex situ* conserved resources and local varieties maintained *in situ* by farmers in traditional farming systems, may be important in the provision of variety sensitive seed relief.

2.1 Materials conserved

The number of genebanks has grown rapidly since the early 1970s, when there were fewer than ten, holding perhaps no more than a half million accessions. A total of more than 1,400 collections are now recorded in FAO's World Information and Early Warning System on Plant Genetic Resources (WIEWS) database. Based on this database, and information provided in Country Reports prepared for FAO, approximately 6.1 million accessions are stored worldwide in *ex situ* germplasm collections, including approximately 527 000 accessions stored in field genebanks. Some very large national collections exist (Table 1) but often the collections are much smaller consisting of only a few hundred or thousand accessions of specific crops. One of the largest groups of collections is that maintained by the IARCs of the Consultative Group on International Agricultural Research (CGIAR). There are now over 600,000 accessions of major food and forage crops (Table 2) in these collections which have been designated as part of the International Network of Base Collections under the Auspices of FAO.

Table 1: *Ex situ* conservation in some of the world's largest national base collections (from FAO, 1996b)

Country and Institute	Accessions	Facilities
China: Institute of Crop Germplasm	300,000	Long-term storage, space available
USA: National Seed Storage Laboratory	268,000	Long-term storage, capacity of 1,000,000 accessions
Russia: VIR	177,680	No long-term facilities
Japan: NIAR	146,091	Long-term facilities
India: NBPGR	144,109	New genebank built for 600,000 accessions
Korea, Republic of: RDA	115,639	Long-term facilities, total capacity 200,000 accessions
Canada: PGRC	100,000	Long-term facilities
Germany: IPK (Institute for Plant Genetics and Crop Plant Research), Gatersleben	103,000	Long-term facilities
Brazil: CENARGEN	60,000	Long-term facilities, capacity for 100,000 accessions
Germany: FAO, Braunschweig	57,000	Long-term facilities
Italy: Bari	55,806	Long-term facilities
Ethiopia: Biodiversity Institute	54,000	Long-term facilities
Hungary: Institute for Agrobotany	45,833	Long-term facilities
Poland: Plant Breeding & Acclimatization Institute	44,883	Long-term facilities
Philippines: NPGRL	32,446	Long-term facilities

Current information in the WIEWS database indicates that over 40% of all accessions in genebanks are cereals. Food legumes are the next largest category constituting about 15% of global collections stored *ex situ*. Vegetables, roots and tubers, fruits, and forages, each account for less than 10% of global collections. Medicinal, spice, aromatic, and ornamental species are rarely found in long-term public collections. Information in the WIEWS database indicates that 48% of accessions for which the type of the material is known are advanced cultivars or breeding lines, while 36% are landraces or old cultivars and about 15% are wild or weedy plants or crop relatives. However, these can only be regarded as approximate estimates as the type of accession is only known for one third of all accessions. More accurate data is available on the CGIAR genebanks collections which appear to be more heavily weighted towards landraces (59% landraces and old cultivars, 14% wild and weedy relatives and 27% advanced cultivars and breeders' lines). In general, the size of the collections of different crops reflects their global importance in production. The largest collections are of wheat, rice, barley and other major staples while crops of local importance (e.g. fonio, bambarra groundnut) tend to be under-represented in genebanks (Padulosi, 1997).

While the centres of diversity of most of the major crops are in developing countries, some of the largest and most extensive collections can be found in developed countries. Thus, there are very large collections in USA, Japan, Canada and Germany (Table 1). Table 3 gives a regional breakdown of accessions.

Table 2: Examples of *ex situ* collections of plant germplasm at selected CGIAR centres (source: SGRP data)

Centre	Crop	Number of Accessions
CIAT	Beans (*Phaseolus*) Cassava (*Manihot*) Tropical forages	41,061 5,985 23,894
CIMMYT	Wheat (including *Triticum*, *Aegilops*) Maize (including Teosinte, *Tripsacum*)	98,905 13,246
CIP	Potato Sweet potato Andean roots and tubers	6,257 6,522 1,132
ICARDA	Barley (including wild *Hordeum*) Wheat (including *Triticum*, *Aegilops*) Chickpea (including wild *Cicer*) Faba bean Lentil (including wild *Lens*)	24,092 30,539 9,974 4,455 7,911
ICRISAT	Sorghum Pearl millet Chickpea Pigeonpea Groundnut Minor millets	35,186 21,191 17,244 12,885 14,957 9,015
IRRI	Rice (*Oryza*)	80,646

Table 3: Genebanks and accessions in *ex situ* collections, by region (FAO, 1996b)

Region	Accessions		Genebanks	
	Number	**%**	**Number**	**%**
Africa	353 523	6	124	10
Latin America and Caribbean	642 405	12	227	17
North America	762 061	14	101	8
Asia	1 533 979	28	293	22
Europe	1 934 574	35	496	38
Near East	327 963	6	67	5
Total	**5 554 505**	**100**	**1 308**	**100**

2.2 Conservation procedures

Where possible, plant genetic resources are conserved as seed in genebanks under long, medium or short term conditions. Seed of many crops can be maintained for very long periods (over 100 years in some cases) provided it is stored at low temperatures (-20 C) and low moisture content. A number of genebanks now have substantial storage facilities which operate at -20 C suitable for long term storage while others maintain material at 4 C for medium term storage. However, there are still many genebanks which do not have any cold storage conditions and have to maintain material at ambient temperatures.

Plant species that are vegetatively propagated, that have long life cycles or produce short-lived (recalcitrant) seed, are commonly maintained in field genebanks. These include crops such as cassava, potato, bananas, plantains, yams, and tree crops such as fruits, coffee, cacao and coconut, which are normally grown in orchards and plantations. Almost every country has at least one field genebank and many countries have several. Although plants in field genebanks can be readily characterized and evaluated, they are also susceptible to loss, due either to pest or disease attack or adverse environmental conditions such as drought, floods, fire, wind, etc. *In vitro* storage is now being developed as an alternative or complementary method. Sixty-three countries report having a tissue culture facility, but it is unlikely that these facilities are all being used for conservation purposes. Information on accessions held *in vitro* is incomplete. Perhaps fewer than 37,600 accessions are being conserved in this manner. The total includes many working collections maintained by plant breeders, as well as those collections established specifically for long-term conservation.

There are now well established guidelines for the maintenance of seeds in genebanks (Ellis et al., 1985). It is recommended that at least 1500-2000 seeds are is maintained of each accession and that whenever germination levels drop below 85% (or for some legume species of crop relatives) steps are taken to regenerate the accession to provide a new stock of high viability. This is an area of genetic resources management that causes substantial problems for many national programmes; even under optimal *ex situ* storage conditions, seed viability will decline, necessitating regeneration in order to replenish stocks. Assuming the regeneration cycle to be 10 years or more on average, one would expect routine annual regeneration needs to amount to less than 10 percent of accessions. However, some 95 percent of countries report a far higher level of regeneration need. Such a situation is an indicator of poor storage conditions, lack of funds or facilities for regeneration, poor management, or a combination of such factors, in many of the world's genebanks.

Wherever possible, genebanks strive to ensure that their holdings are duplicated in another institution. This provides security in the case of loss of material in the genebank through some accident or other cause. In practice, duplication of the world's genetic resources holdings is very incomplete since it requires substantial resources to ensure that sufficient quantities of each accession are multiplied, packed and shipped to an appropriate safety repository. Nonetheless, a number of genebanks, particularly those of the CGIAR are making significant efforts to ensure that safety duplication is undertaken. In addition to this deliberate effort to ensure the safety of whole collections, there is extensive casual duplication of accessions as a result of transfers between genebanks and different breeders or research groups over the last 2-3 decades.

2.3 Information management

It has always been recognized by those involved in conservation of plant genetic resources that information on the resources is as essential as the resources themselves. The most important information on any accession would include such items as taxonomic status, crop and variety name, original source (location) of the material, genebank accession number and other genebank management data. This provides the basic information on what an accession is and where it comes from that would be needed in any attempt to locate potentially useful material for disaster relief (see 3.3 below). Other important information would be that on the agro-morphological characteristics of the accession including its morphology, yield, cropping characteristics and resistance to pests, diseases and stresses. These data are normally described as characterization and evaluation data. A good overview of current genetic resources information management practices and procedures is provided by Painting et al. (1993) and IPGRI (1994).

It has to be accepted that many of the world's *ex situ* collections are insufficiently and poorly documented. While some countries have fully computerized documentation systems and reasonably complete accession data (most European Countries, the USA, Canada, Australia, Japan, China, India, Brazil, Ethiopia, and Kenya), many countries have only partial computerization of documentation systems and much more limited data on accessions. In countries with decentralized *ex situ* germplasm collections, such as several in Western Europe, databases are maintained by the individual institutes. However, centralized documentation systems exist, e.g. in USA (the GRIN system) and in Germany at the German Information Centre for Genetic Resources. Despite the substantial progress that has been made in recent years, many countries simply lack information on the accessions in their own collections and many genetic resources workers would emphasize the need for integrated, compatible systems which allow for easy exchange of information.

The Internet facility allows direct access to the information on plant genetic resources on international level (SINGER, WIEWS, ECP/GR platform, IPGRI Crop Directories) as well as to the national documentation systems (GRIN in USA, GENRES in Germany, Nordic Gene Bank in Scandinavia and many others). Internet addresses of many of these can be found via WIEWS and some are listed in Annex 1.

The System-wide Information Network for Genetic Resources (SINGER) is the genetic resources information exchange network of the CGIAR. It provides common access to information concerning the collections of genetic resources held by the CGIAR Centres. SINGER links the genetic resources databases of the CGIAR Centres and allows searches for information relating to the identity, origin, characteristics and distribution of the genetic resources in the individual Centre collections and access to further specific data on the collections, e.g. crop characterization data. Similar activities can be carried out using GRIN and it is likely that Internet access to an increasing number of genebank holdings will become available in the next few years.

The WIEWS system provides information on national programmes on conservation and utilization of plant genetic resources for food and agriculture and consists of a number of databases and information retrieval software including geographical presentation (maps) of location of *ex situ* plant collections. It provides a register of institutions involved in plant genetic resources conservation and utilization with a list of key scientists in each institution and a list of crops and activities dealt with. For each country, all institutions are given a series of numbers that hierarchically identifies them in the structure of the national plant genetic resources programme. The *ex situ* collections database provide a summary record of each genebank's (collection's) holdings. The data includes species name and number of accessions for this species, type of material held (wild, land race, advanced cultivar, breeder's material, mutants etc.), geographical distribution of accessions, place of safety duplication. For each collection the technical parameters of storage are recorded. The "Early Warning" database is being developed and will consist of reports on the possible genetic erosion in the natural environment and *ex situ* plant collections.

Information on more recently collected material is often substantially better than that on material collected early in the work on conservation. FAO and IPGRI both provide information on the kind of information that should be obtained as part of germplasm collecting work (FAO, 1994). Those involved in disaster relief activities where material is obtained directly from farmers or specific localities may wish to consider using some of these procedures to ensure that they have the right information on the materials they obtain.

2.4 Availability and Distribution

Most genebanks would place considerable emphasis on the importance of making material available to potential users of plant genetic resources for crop improvement and other purposes. Distribution to workers within the same country is often a key activity of many genebanks and the importance of international distribution is also recognized. Availability has 2 aspects; the procedures that need to be followed for distribution and exchange of accessions and the physical availability of genebank accessions.

With respect to international availability, though there are important differences, no country can claim self-sufficient in the genetic resources needed to sustain and improve agricultural production. According to recent studies, any region of the world is dependent on genetic material, which originated in other regions, for over 50% of its basic food production, and, for several regions of the world, such dependency is close to 100%. Because of such global interdependency, the access to plant genetic resources is an important component of some of the international agreements and Conventions, particularly the Convention on Biological Diversity (CBD) and the International Undertaking on Plant Genetic Resources.

The International Undertaking and CBD express the objectives and commitments of the international community on issues relating to the conservation of, access to, and sustainable utilization of plant genetic resources. The TRIPs agreement contains the internationally agreed minimum standards for Intellectual Property Rights (IPRs), including on plant varieties, through patents, or *sui generis* systems.

Another aspect of international concern is that of plant health status. Trade in plants and plant products and the exchange of germplasm is conducted within the framework of the International Plant Protection Convention (IPPC). The IPPC, deposited with the FAO, is recognized as the primary instrument for international co-operation in the protection of plant resources from injurious pests and the harmonization of phytosanitary measures affecting trade. The exchange of germplasm internationally for research, conservation and basic plant breeding purposes, including material for biotechnology, requires specific phytosanitary information before the exchange can take place. Failure to follow these procedures can result in the introduction of new pests into countries or regions, and can ultimately result in epidemics that threaten food security and biodiversity. The IPPC Secretariat facilitates and assists in identifying technically sound practices that safeguard against the introduction and establishment of unwanted pests. It is for these reasons that most countries have sound, and often detailed, regulations on phytosanitary measures.

Often the quantities of any accession that may be available to the interested user is very small. Genebanks normally seek to be able to distribute quantities which can range from 100 seeds (for large seeded species) to 100 gm. These quantities are so small than one generation of further multiplication will usually be needed for many purposes such as multilocational trials and entry into breeding and evaluation programmes of agricultural institutes. Even these amounts might often not be available for some accessions. As noted above, there are also problems in availability that reflect the backlog of regeneration that many genebanks face and a special multiplication might be required before material can be made available to a potential user.

2.5 Institutions involved

Country based efforts form the central element in the global effort to conserve and use plant genetic resources. The work is often undertaken by more or less formally constituted national genetic resources programmes. The purpose of national PGRFA programmes is to contribute to national development, food security, sustainable agriculture and the maintenance of biodiversity through the conservation and utilization of PGRFA. The main functions are: development of national polices and strategies, co-ordination of national activities involving all stakeholders, providing linkages and providing the building blocks of regional and international collaboration.

Activities of a national PGRFA programme can be very wide ranging. They can include inventorying, exploration and collecting of germplasm, training and capacity building, *in situ* and *ex situ* conservation, research, characterization and evaluation, fund raising, genetic enhancement, development of legislation, crop improvement, regulation of access and exchange of genetic resources, seed and variety production and distribution, documentation and dissemination of information and public awareness raising.

It follows that a very diverse range of institutions can be involved in national plant genetic resources programmes and their organization and linkages is subject to extensive variation. Ministries and government departments often play a key role and the Ministry of Agriculture is often a central coordinating body working together with ministries responsible for forestry, environment, natural resources, science and technology, planning and development. A number of national agricultural research institutes are often involved in national programme activities together with Universities, Technical Colleges, Botanic Gardens and other research and development agencies. Important roles can also be played by NGOs, farmers' organizations, women's groups and companies in the private sector concerned with agriculture and forestry.

Both regional and national organizations also play an important part in conservation work. Regional organizations can include research institutions such as CATIE which is based in Costa Rica or AVRDC based in Taiwan and concerned with vegetable production. A number of regional networks have been established to support linkages between national programmes. These exist in all parts of the world and bring together national programmes for specific activities concerned with information management, germplasm exchange and distribution and research and development. They include for example WANANET (operating in West Asia and North Africa), REMERFI, REDAFIT and TROPIGEN (working in S and Central America), ASARECA in Africa and APAARI in Asia (Riley et al., 1995).

The CGIAR system and its associated centres plays a special role at the international level. Established in 1971, the CGIAR is an informal association of 58 public and private sector members that supports a network of sixteen international agricultural research centres. The mission of the CGIAR is to contribute, through its research, to promoting sustainable agriculture for food security in the developing countries. As noted above, CGIAR centres hold in trust for the future one of the world's largest *ex situ* collections of plant genetic resources, containing over 600,000 accessions of more than 3,000 crop, forage, and pasture species. The collection includes improved varieties and, in substantial measure, the wild species from which those varieties were created. These materials are freely available to researchers around the world so that new gene combinations can be brought to bear on current problems. The CGIAR was the first to place its collections under the auspices of FAO as the basis of an international network of *ex situ* collections. The accessions in these collections will be released to bona fide users, through Material Transfer Agreements (MTAs) that require that no IPRs can be sought over the material. These materials are available for emergency situations and disasters.

FAO has a major role in global conservation work. With a mandate to raise levels of nutrition and standards of living, to improve agricultural productivity, and to better the conditions of rural populations, the organization gives highest priority to food security - access of all people at all times to food they need for an active and healthy life. Recognizing the importance of genetic resources for food production and agriculture development, FAO Conference in 1983 established the Commission on Plant Genetic Resources (now the Commission on Genetic Resources for Food and Agriculture - CGRFA), through which countries, inter alia, monitor and oversee the development of a Global System for the conservation and sustainable utilization of PGRFA, including the International Undertaking and the Global Plan of Action. The International Undertaking is currently being renegotiated to bring it in full harmony with CBD, especially its provisions on access with CBD principles. These negotiations are premised on the importance of ensuring, at least for a number of crops, a multilateral system that facilitate access to and exchange of PGRFA. In addition, FAO maintains a number of important activities such as the development and maintenance of WIEWS.

2.6 *Ex situ* and *in situ* conservation

While plant genetic resources activities have traditionally been conserved with ex situ conservation, the importance of supporting *in situ* conservation activities is now increasingly recognized. Formal *in situ* conservation activities are being implemented by a number of countries both for wild relatives of crop plants (e.g. Ertug and Tan, 1997) and for local crop varieties (Worede, 1997; Jarvis and Hodgkin, 1999). The role that these activities could play in disaster preparedness and relief has not been addressed to date and needs further consideration. However, one important feature of these activities is that they retain the full diversity and integrity of local cultivars with, hopefully, sufficient adaptive variation to allow their continued evolution and adaptation to changing conditions. Local cultivars in some collections have been separated into a number of different accessions comprising the different types found in them. While this should ensure maintenance of all the diversity present in the local variety it may make reconstituting a local cultivar for use in disaster relief more difficult.

3. PGR AND DISASTER PREPAREDNESS AND RESPONSE

Plant genetic resources have considerable potential to contribute to the provision of appropriate seeds (or other planting materials) following disasters. In this section we consider some of the key issues involved in making this contribution from the perspective of identifying the range of resources available, determining the issues involved in finding those most appropriate for specific situations (and the capacities to do this) and considering how to use them.

However, it is also important to note and accept the limitations of seed provision as an appropriate intervention following a disaster. As other have noted (e.g. ODI,1996) there are situations where provision of seed is undesirable (e.g. conditions are not yet suitable for a return to production), or unnecessary (e.g. secret seed stocks exist). There are also occasions where other needs are paramount such as the need to earn off farm income or where other materials (tools etc) are needed more.

3.1 Identifying material needs

Where seed provision can make an identified contribution, a primary consideration should be an understanding of the variety needs of the affected communities. Important components in the analysis of these needs include a consideration of the crops required, the phenological and ecological characteristics required of the crops in the affected areas (daylength adaptation, cold hardiness etc). A knowledge is also required of the characteristics of the production systems and the current status of the various socio-economic factors that are important in the management of production and of seed supply systems. An area of particular concern is the extent to which previous production materials and systems are still appropriate in the wake of the disaster.

There is a common presumption that the material required should match as closely as possible that which has been destroyed in the disaster. This may be generally true and it is certainly likely to be the basis of any disaster preparedness plans. However, it may not always be the case. For example, there may be a need for short season (or even off-season) varieties to ensure the rapid production of some food under abnormal conditions. There may also be a need for material that produces well under low input agriculture for situations where fertilizers are lacking or normal irrigation systems have been destroyed. Finally, the social or economic situation may have changed to such an extent that new production systems are developing for which new varieties are needed. These issues are discussed more fully in Richards and Ruivenkamp (1997). Plant genetic resources and information can still be deployed to meet these different needs but the procedure for identifying the optimum resources has to take account of these specific factors.

3.2 Local sources of material

Other papers prepared for this meeting will deal more fully with issues of supporting local seed systems and locating local materials from the production system (e.g. Farmer Seeds Systems in Disaster Situations; Seed Stocks and Seed Multiplication in Crisis). From a genetic resources perspective, it is worth noting that an increasing number of countries are concerned to develop effective systems for conserving local varieties *in situ.* The form that these will take will vary considerably and is currently the source of much debate. However, they may well provide sources of useful material with proven adaptation to the environments in which they are grown and sufficient variability to adapt to changing environments and production conditions (see also 2.6, above).

It is also worth considering whether the information contained within genetic resources conservation systems can be used to help locate potentially useful materials. The results of many years of collecting, characterization and evaluation of plant genetic resources may well provide information on where appropriate types might still be found and what characteristics should be looked for.

One problem with respect to the potential use of material conserved *in situ* is that information management systems, which parallel those developed for *ex situ* conserved materials, have yet to be developed. There is a critical need for NARS to develop databases on plant genetic resources of traditional varieties grown in their countries. Inventories of plant germplasm in production systems which include information on agro-ecological production conditions, crop characteristics and relevant socio-economic data would be extremely valuable and are important elements both of CBD and of the Global Plan. A regional Seed Security Network could serve as a repository for this information gathered by NARS in the region and could be tapped when needed for information on stocks that need to be replaced in areas affected by disaster in other locations in the region.

3.3 Information availability and capacities

The information on plant genetic resources that can be used when identifying optimum resources for relief situations depends very much on the quality of documentation systems in genebanks. Data are basically divided into five categories:

- Passport data which includes basic information on a sample conserved in a genebank - name of a variety or local name of a cultivar, botanical characteristic, country of origin etc.;

- Management data which provides the basis for the management of a sample in the genebank in order to assist multiplication and regeneration procedures (germination percentage, type of regeneration procedure);

- Characterization data on attributes which are highly heritable and can be seen by eye and are equally expressed in all environments (shapes, colours etc);

- Evaluation data which includes information on agronomic characteristics which are susceptible to environment differences (e.g. yield, agronomic performance and stress susceptibilities); and

- Environment site data which includes environmental and site-specific parameters when characterisation and evaluation trials are held. Site information for the area from which the material was originally collected is also in this group.

Unfortunately information on accessions is seldom complete and it is not uncommon to find that accessions possess only limited passport and management data with little or no characterization or evaluation data. One problem in attempting to use plant genetic resources in disaster relief activities may be the limited data available to identify the most useful accessions. However, in a number of genebank collections a substantial fraction of the accessions possess sufficient information to make searches to find suitable material well worth undertaking providing the limitations of the information are accepted.

The first line of investigation may well be to identify accessions of relevant crops which originate from the country of concern or from neighbouring countries sharing similar production systems and production environments. This very simple approach was the one used to identify Somali maize and sorghum materials held in Kenya which could then be used in the CINS/IPGRI initiative. A similar approach was used for the Rwanda Seeds of Hope initiative where CIAT was able to identify accessions obtained from Rwanda which could be provided in case of need. Large international genebanks often have substantial collections from many countries and the SINGER database or the USA GRIN database can be used to identify such materials rapidly and easily. As an example, Table 4 lists the major collections of sorghum and Table 5 the countries from which the material held by ICRISAT (the world's largest single collection) was obtained.

Where passport data includes latitude and longitude, direct mapping of the original site of collection of accessions is possible. This is often the case for only a small proportion (20-50%) of accessions. However, sufficient site information is often available to allow interested users to map accessions and calculate appropriate latitude and longitude data. For the world's collections of wild wheats, this was possible for 70% of the accessions (Hodgkin et al., 1992).

Table 4: Major collections of Sorghum in different countries

Genebank	Country	Accession	
		Number	%
ICRISAT	CGIAR	35,186	21
NSSL	USA	18,971	11
S-9	USA	15,434	9
VIR	Russia	9,892	6
PGRC-E	Ethiopia	7,260	4
DTCP-CSIRO	Australia	7,000	4
CNPMS	Brazil	6,589	4
ICGR-CAAS	China	5,263	3
USDA	Puerto Rico	4,000	2
ASP	Yemen	4,000	2
INTA-EEAMANF	Argentina	3,900	2
ARC-Sudan	Sudan	3,145	2
CIFAP-MEX	Mexico	3,000	2
NRC-Sorghum	India	2,767	2
SAARI	Uganda	2,635	2
NIAR	Japan	2,583	2
INIA-Iguala	Mexico	2,500	1
NPGRL	Philippines	2,285	1
CENARGEN	Brazil	2,276	1
Others		29,864	18
Total		**168,550**	**100**

Geographic information systems (GIS) analysis can be combined with these simple geographic approaches to increase the specificity of the search process. In this way it should be possible to select sets of material which are likely to have more relevant characteristics. For example, location information (from passport data) can be combined with information on climate or soils in the affected areas to locate material adapted for specific production environments (e.g. with specific altitude, temperature, daylength, soil or other characteristics). These techniques are being increasingly used to analyse the extent and distribution of genetic diversity and their application in this context would now seem very feasible (for a recent description of the use of GIS techniques for germplasm management activities see Guarino et al., 1997).

Mapping techniques can be complemented by a wide range of bibliographic information on socio-economic aspects and on cultural preferences and other characteristics. These can also be combined with any agronomic data already maintained by the genebanks on the accessions themselves. Often cultural and socio-economic data will need to be collated from other sources since these are not areas in which gene banks have traditionally maintained much information.

Table 5: Country of origin of ICRISAT's sorghum collections according to SINGER (only countries providing over 100 accessions are listed)

Country	No. of accessions	Country	No. of accessions
India	6092	China	380
Ethiopia	4372	Lebanon	360
Cameroon	2479	Zambia	341
Sudan	2456	Togo	294
Yemen, Republic of	2129	Rwanda	290
United States of America	2065	Lesotho	269
Uganda	1730	Central African Republic	249
Nigeria	1678	Senegal	239
Zimbabwe	1599	Botswana	218
Kenya	971	Swaziland	202
South Africa	902	Benin	197
Tanzania	701	Chad	187
Mali	701	Namibia	181
Burkina Faso	546	Venezuela	173
Somalia	445	Ghana	146
Niger	409	Burundi	136
Malawi	403	Japan	108
Union of Soviet Socialist Republics (former)	395	Sierra Leone	107
		Others	1092

There is clearly a considerable potential to use a range of integrated sources of information available from plant genetic resources databases and other systems to identify material that might be of relevance in responding to seed needs following disasters. There are also limitations which reflect partly the availability of information on genebank accessions and partly the lack of experience that exists with respect to the best ways of using the information that is available. This is an area where coordinated investigations on a case study basis would be extremely rewarding in determining the potential and pitfalls of these information techniques as part of search systems for suitable materials.

3.4 Institutional capacities

Nearly all countries possess a plant genetic resources programme of some type that provides an institutional framework for providing relevant genetic resources and information for disaster relief. Of course, in many situations the institutions may themselves be substantially affected by the disaster and find it difficult or impossible to respond. However it is likely that there will still be some people who have substantial knowledge that is relevant to the identification of optimum materials and their involvement is essential.

As described in Section 2, national genetic resources programmes can comprise a very large number of different organizations and groups. Obviously, they may also be very small and consist of a few breeders or conservers working on a range of crops with extremely limited resources. However, there are potentially a wide range of organizations that could be consulted and mobilized to provide relevant information on genetic resources. Institutions directly involved in conserving materials and in using them in crop improvement work can provide information on their holdings and often on holdings of others (such as CG Centres or genebanks in other countries). They may also provide access to sources of material in national breeding programmes or other research centres. Information on ecological characteristics, on soils, geology, climate will come from other institutions and ministries and can be used to identify areas in which relevant adapted material can be found outside the disaster area. NGOs and farmers groups will provide information on locally important characteristics that need to be considered in locating appropriate materials.

Both regional and international institutes and organizations can also supply relevant information to help locate suitable resources. These usually already have some linkages to national programmes, especially such as IPGRI and UN agencies such as FAO, which maintain close links with many national programme personnel. Networks that are concerned with information and material exchange can also provide up to date information on relevant germplasm resources and the information available on it.

While there are many institutions with potential to provide useful and relevant information and materials, there are likely to be substantial problems in realizing their full potential. Resources to respond to needs will be often be extremely limited. Locating relevant information or materials will often be impossible unless additional resources can be made available to support searches and evaluation of both the information and the material. The required collaboration between different organizations and groups may not exist and there may be an incomplete appreciation of the needs and how they can best be met. Discussion of the practical implications of using plant genetic resources to assist in disaster relief at national and regional levels within the context of implementing the Global Plan is clearly needed.

4. BUILDING IMPROVED RESPONSE SYSTEMS AND CAPACITIES

A basic objective of introducing the genetic resources component in seed relief work is to optimize the provision of material through the best use of all the different information and seed resources that are available. It is likely that there are a number of situations where genetic resources, the information on them and the institutions and people involved in their conservation and use can make a material contribution to providing appropriate seed materials following different types of disaster. Institutional support is needed to strengthen linkages between those involved in provision of seeds to disaster-affected agricultural systems at national, regional and international levels. Key elements that should be sought through this process and that would maximize the contribution of genetic resources include:

4.1 Institutional considerations

Links are clearly needed between those involved in provision of seed in disaster situation and those involved in management of genetic resources. These links are needed at national, regional and international levels. The national programmes in many countries are complex and dispersed involving many different institutions. In addition there are many regional and international organizations also involved in genetic resources management. Appropriate focal points might need to be identified who could help to coordinate planning and development of the work needed and provide the necessary liaison between different actors. At the same time it is necessary to recognize that resources will be needed to support any new initiatives and develop an effective programme. A valuable contribution would be to strengthen regional pgr networks and enhance their capacity to manage information on genetic resources availability in *ex situ* collections for seed relief programmes.

Activities undertaken by national programmes under their CBD obligations or in support of the implementation of the Global Plan can materially contribute to the improved identification of appropriate material for seed provision following disasters. A key element would be improved inventories of local variety use at country level.

4.2 Material availability

Even when useful material can be found in genebanks the amounts of seed available are always likely to be small. Ways in which this problem might be addressed are needed. It might be possible to develop reserve stocks of key adapted types in some genebanks or to strengthen on farm conservation systems. Links are also needed between genebanks and those routinely involved in multiplying seeds (seed companies) so as to benefit from their experience and resources. One specific problem that may need to be addressed by some genebanks is that of maintaining local cultivars in their original state and not separating them into separate components.

4.3 Information

Information on genebank holdings and the systems and procedures developed for managing and using such information may prove to be one of the greatest contributions that the genetic resources community can make. However further investigations are needed to determine how the information can best be used, what its strengths and weakness are in the context of the activities envisaged and what improved systems could be developed. This might best be done through collaborative studies by those involved in seed relief and by relevant PGR institutions to develop and test appropriate ways of using information on pgr to assist in disaster relief and locate useful materials.

5. ACKNOWLEDGEMENTS

The authors thank many colleagues for their helpful inputs into the preparation of this paper,, particularly, Victor Bushamuka, Nina Dudnik, Jan Engels, Pablo Eyzaguirre, Bill Fiebig, Dave Nowell, and Jurek Serwinski.

6. REFERENCES

Ellis, R.H., T.D. Hong and E.H. Roberts, 1985. Handbook of seed technology for genebanks, Vols. I and II, IPGRI, Rome.

Ertug, A. and A. Tan, 1997. *In situ* conservation of genetic diversity in Turkey. In: N. Maxted, B.V. Ford Lloyd and J.G. Hawkes, Plant genetic conservation: the *in situ* approach. Chapman & Hall, UK, pp. 254-262.

FAO, 1994. International code of conduct for plant germplasm collecting, FAO, Rome.

FAO, 1996a. Global Plan of Action for the Conservation and Sustainable Utilization of Plant Genetic Resources for Food and Agriculture. FAO, Rome, pp. 63.

FAO, 1996b. Report on the State of the World's Plant Genetic Resources for Food and Agriculture. FAO, Rome, pp. 75.

Guarino, L., N. Maxted & M. Sawkins (1997) "Analysis of geo-referenced data and the conservation and use of plant genetic resources." Linking Genetics and Geography: Emerging Strategies for Managing Crop Biodiversity. ASA/CSSA/SSSA Annual Meetings. 26-31 October 1997. Anaheim, California.

Hodgkin, T., Adham, Y.J. and Powell, K.S. (1992) A preliminary survey of wild *Triticum* and *Aegilops* species in the world's genebanks. Hereditas 116: 155-162..

IPGRI, 1994, Genebank Standards, IPGRI, Rome, pp. 13.

IPGRI, 1998. Re-introduction of germplasm for diversification of the genetic base of the Somali agricultural production system. IPGRI Report.

Jarvis, D. and T. Hodgkin, 1999. Farmer decision making and genetic diversity: Linking multi-disciplinary research and implementation on form. In: S. Brush (eds.) Genes in the Field, C.R.C. Lewis (in press).

ODI, 1996. Seed provision during and after emergencies. Relief and rehabilitation network. Good Practice Review 4, ODI, UK.

Padulosi, S., 1997. Underutilized species: status and initiatives aimed at better conservation and use. Congreso Nacional de la Sociedad Espanola de Genetica, Valencia, Spain, 11-13 September 1997. (in press)

Painting, K.A., M.C. Perry, R.A. Denning and W.G. Ayad, 1993. Guidebook for genetic resources documentation. IPGRI, Rome, pp. 295.

Richards, P. and G. Ruivenkamp, 1997. Seeds and survival: crop genetic resources in reconstruction in Africa. IPGRI, Rome, pp. 61.

Riley, K,W., Zhou Ming De and V.R. Rao, 1995. Networks for the effective management and use of plant genetic resources in Asia, Pacific and Oceania. Paper presented at meeting on Coordination and Evaluation of Network of PGR at National, Regional and International Level, Yogyakarta, September, 1996.

Sperling, L. 1995a. Emergency Relief in Rwanda: Assesssment of bean and seed use. Summary report. SOH Assessment Document 1; Impact of war on plant genetic resources: bean varietal Assesssment in Rwanda. Summary report. SOH Assesssment Document II.

Worede, M. 1997. Ethiopian *in situ* conservation. In: N. Maxted, B.V. Ford Lloyd and J.G. Hawkes, Plant genetic conservation: the *in situ* approach. Chapman & Hall, UK, pp. 290-301.

Annex 1: Information sets and databases accessible and searchable online through the Internet

n.b.: URLs were correct on 15 September 1998

Name	**Description**	*URL*	*Availability in other format*
Central Crop Databases, Directory of European PGR Collections (ECP/GR)	**The European Information Platform on Crop Genetic Resources, established under the umbrella of the European Cooperative Programme for Crop Genetic Resources Networks (ECP/GR) provides online access to regional/global crop specific databases (genetic resources conserved in genebanks throughout the Region) and contact information for institutions which are active in crop genetic resources conservation and utilization. Several components of this information platform are still under development. In particular the number of European Crop Databases to which on-line access is given, will increase considerably in the near future.**	**http://www.cgiar.org/ecpgr/platform/Index.htm**	**Printed publication: Frison, E.A. and J. Serwinski 1995**
Directory of Arid Land Research Organizations (FAO)	**Contains contact details, research programmes, scope of interest, facilities, etc. of arid land research organizations world-wide. It provides a list of countries and an index of the institutes. Based on the 4th edition (1995) of the Directory.**	**http://www.fao.org/WAICENT/FAOINFO/FORESTRY/ARIDLAND/ARIDLAND.HTM**	**Printed publication: Waser, K.V. and B.S. Hutchinson, 1995**
Directory of Germplasm Collections, IPGRI	**Maintained by the International Plant Genetic Resources Institute IPGRI. The Directory provides summary information on ex situ germplasm collections worldwide. The data include address information on organizations holding germplasm and summary information on the type of germplasm that is maintained such as, species names, number of accessions per species, type of accessions, etc. Currently summary information on more than 5 million accessions worldwide is available.** **Searches can be defined on germplasm characteristics (taxon, type of material, country of origin, number of accessions) and/or holding location (name institute, city, country). Addititonal information, such as safety-duplication status or parameters on seed strage can be obtained from IPGRI.**	**http://www.cgiar.org/ipgri/doc/dbintro.htm**	**Printed Directories of Germplasm Collections (At present 16 directories covering all major agricultural crops)**
ECOCROP 1 (FAO) (Test version)	**The ECOCROP1 database was developed as a tool to identify plant species (about 1700) for given environments and uses. It can be used as a library of crop environmental requirements and it can provide plant species attribute files on crop environmental requirements to be compared with soil and climate maps in Agro-ecological zoning (AEZ) databases or Geographical Information System (GIS) map-based display. Searchable by crop, soil and climate requirement.** **Future developments: expanded information on crop husbandry, seed sources, plant genetic resources.**	**http://www.fao.org/waicent/FaoInfo/Agricult/AGL/AGLS/eco1.HTM**	

Ecosys	"Beckstrom-Sternberg, Stephen M. and James A. Duke. The Ecosys Database. Http://probe.nalusda.gov:8300/cgi-bin/browse/ecosys." 885 records of plant ecological ranges including ranges for temperature, rainfall and soil pH. Searchable by taxon, genus, family.	www.ars-grin.gov/duke/	
Famine Foods DataBase	List of plants that are not normally considered as crops and their uses as food. Provides species lists by family or species name and gives information about a species' use and its common name.	http://www.hort.purdue.edu/newcrop/FamineFoods/	
GRIN Germplasm Resources Information Network	Maintained by the United States Department of Agriculture (USDA), Agricultural Research Service's (ARS) National Plant Germplasm System (NPGS).The GRIN database contains information on all genetic resources preserved by NPGS, including accessions of both domestic and foreign origin (currently more than 450 000 accessions of economically important vascular plants). Though the emphasis is on major, minor, or potential crops and their wild and weedy relatives, many other categories of plants are represented including ornamentals and, more recently, some rare and endangered plants. Can be searchec by accession, crop science registration, plant variety protection and taxonomic queries(see GRIN taxonomy) and crop lists and their descriptors	http://www.ars-grin.gov/npgs/searchgrin.html	
SINGER CGIAR's System-wide information Network for Genetic Resources	SINGER links the genetic resources databases of the CGIAR Centres and allows searches for information relating to the identity, origin, characteristics and distribution of the genetic resources in the individual Centre collections and access to further specific data on the collections, eg. crop characterisation data. There are a variety of search possibilities: Accession level and crop-based searching categories, predefined searches, characterization searches. Continuously updates by the CGIAR centers.	http://noc1.cgiar.org/	CD-ROM, data as of 1st April 1997. Free of charge, available from the SINGER coordinator, International Plant Genetic Resources Institute, Via delle Sette Chiese 142, 00145 Rome, Italy, tel (+39)6-518921, fax (+39)6-5750309
WIEWS (FAO) World Information and Early Warning system on Plant Genetic Resources	The system provides information about national programmes on conservation and utilization of plant genetic resources for food and agriculture and consists of a number of databases and information retrieval software. Country Profile module: describes the structure of national Programmes Ex-situ Collections database: summary record of each genebank's (collection's) holdings; Early Warning database: under development. Will consist of reports on the possible genetic erosion in the natural environment and ex situ collections. The system can be searched by institute, genebank, species, contact people and crop activities	http://web.apps.fao.org:8080/wiews.new/	

REGULATORY ASPECTS OF SEED SECURITY

Niels P. Louwaars
CPRO-DLO, Wageningen, The Netherlands
and
Dr. R. Tripp
ODI, London, UK.

Summary

Seed security is a complex concept, which deals with both the sustainability of seed supply and with (preparedness for) emergency seed provision after disaster. Regulations are in place in many countries in order to support the development of a formal seed system. These include seed laws (seed councils, variety release and seed certification), phytosanitary laws and intellectual property rights systems. Since most farmers in developing countries depend on farm saved and locally obtained seeds more and more organisations deal with supporting such diversified seed systems. It is important from a seed security perspective that regulations do not unnecessarily interfere with these local seed systems.Secondly, when a natural or man-made disaster strikes a particular region, seed is increasingly often included in emergency packages that follow (or replace) food aid. The regulations may hinder both procurement and distribution of locally adapted seeds.This paper intends to present an overview of the regulatory issues involved in seed security programmes and suggests measures to support the development of regulations that support seed security both at a national and a regional level.

1. INTRODUCTION

1.1 Seed systems

Seed is a vital input in crop production. In any kind of agriculture, seeds and vegetative planting materials are a basic input. The seed carries the physiological basis for vigorous crop development, and the genetic basis for the adaptation of the crop to both growing conditions and product preferences. Seed and the accompanying knowledge systems are valued in all farming communities as a major cultural element. Ever since seed is considered an important vehicle to extend intensified production techniques in developing countries, the supply system has received considerable attention. These formal systems, designed along the lines of western organisation patterns of seed supply, have replaced the age-old local seed supply systems in particular regions and crops. Limitations of these systems have led to the development of the concept of integrated seed supply (Louwaars 1994).

The two basic seed supply systems are broadly defined as: Formal seed supply: "the chain of activities from breeding to marketing/distribution, run by specialised 'seedsmen' and supported by well defined rules and procedures, that supplies seeds to farmers with some level of quality guarantee. These systems are rather uniform in time and space." Thomson (1979) and Wellving (1984) have described the formal system in some detail. Local seed supply: "activities within the farming community that aim at ensuring the availability of seed for the next planting. These systems are heterogeneous in space and flexible in time." Local seed supply systems have been summarised and discussed by Almekinders (et al, 1994). The objective of both systems is to make sufficient quantities of seed of optimum quality available for each planting season.

A major factor, distinguishing formal and local seed supply is that the former is vertically organised, whereas the latter system can be considered horizontal or cyclic. In formal seed supply, activities follow each other. This system is rightfully compared with a chain. The main links are plant breeding, seed multiplication and seed distribution. Since a chain is as strong as its weakest link, these links have to be developed in harmony with each other. There is no point in producing large quantities of seed without a well-developed distribution system; an efficient seed production and marketing system cannot survive without the regular supply of new varieties and breeder's seed.

Local seed supply consists of basically the same components: selection, production and diffusion but contrary to the vertically organised formal systems, they are in most cases part of common crop production and as such organised in a horizontal manner. Seed production is the focal point here; selection and diffusion are not necessarily given the same emphasis every seed production cycle. The horizontal pattern of local seed supply systems seems to imply that they are more sustainable. The formal seed system is important for a very limited number of crops in developing countries. Public formal systems commonly deal with the most important (field) crops only, such as rice, wheat, maize, cowpeas. Commercial formal seed systems concentrate on high value seeds such as (hybrid) maize, pearl millet (India) and vegetable seeds. Local seed supply caters for over 90% of the total seed need in almost every developing country. A typical aspect of local seed systems is that they may maintain a wide diversity within and among varieties or landraces as a response to diverse ecosystems and local markets. Formal systems can by the very nature of centralised organisation carry only one or few varieties. Genetic resource specialists have started to gain an interest in supporting local seed systems as part of their in situ strategies for the conservation of agro-biodiversity. Heterogeneous farming conditions both in space and time are likely to result in heterogeneous populations of a crop. Especially in such conditions, landraces provide a balance between high yield potential and high yield security.

Especially in (near) subsistence farming systems where seed security is likely to be challenged regularly, the local system is the predominant supplier of seeds.

1.2 The choice of varieties

The farmer's choice to grow certain crops and particular varieties of these crops is based on a complex set of decisions. These are based on needs and expectations of the individual, the farm household and the community. The needs can be summarised as economic and cultural. The expectations are basically the level at which consumption of the household should be met with regard to food, feed and fibre on the one hand, and the market expectations for surpluses or crops grown particularly for the market on the other. The ecological conditions in the farming system form the bottom line for crop production. A good understanding of the choices that farmers make with regard to varieties is very important both in seed system support and in emergency seed provision.

Economic Aspects

The choice of crops depend on such considerations as: should a farmer produce most of the products that the household requires or can he afford to concentrate on market production and subsequent market purchases for agricultural products. This issue is linked with attainable yield and with differences in risk that farmers may accept and with product quality. Yield depends mainly on ecological and farm management factors, combined with the way that the crop may respond to such factors (see 'ecological aspects' below). Varieties have to fit in the farming system, where particular crops have multiple uses. Modern varieties are often bred with a very limited number of objectives, concentrating on grain yield and its components, such as harvest index and stress tolerance/resistance. In most cereal based farming systems in the WANA region, for example, plant and animal production are integrated. Straw yield and straw quality may in many cases be at least as important as grain yield. Short straw varieties bred for a high harvest index, may thus not become very popular. Risk is a factor that is often overlooked in formal research, but a major factor for many farmers, especially those farming in ecologically marginal conditions. When farming is carried out mostly for subsistence, farmers can not afford to take risks without jeopardising the mere existence of those depending on him. In the choice of varieties considerations of risk may prevail over yield expectations, which could be an explanation of less than expected adoption of new varieties in many regions of the world. Uniform and so-called high yielding varieties that exhibit a higher variation in yield between years may not be useful for such farmers even when average yields over a large number of years are proven for these conditions.

Cultural Aspects

Apart from purely economic aspects, also cultural aspects play a role in the choice of varieties. Old varieties are considered a cultural heritage by many communities and grown 'out of tradition'. The varieties, grown and selected by many generations are then to be considered an important aspect of the cultural identity of the farmer. A more practical aspect of culture in variety choice is the sometimes very locality-specific consumption values that crops have to meet. Often farmers produce modern varieties for sales in the market alongside local varieties for home consumption. Finally, local post harvest and processing methods may have an important influence on the choice of varieties. Because food processing is in most cases a woman's job, this introduces an important gender aspect in the choice of varieties.

Ecological aspects

The response of varieties to ecological factors and in particular the abiotic and biotic stresses determine the output with regard to both yield and quality factors of the product. Crops differ in the genotype x environment interaction for important traits like yield, but a generalised observation is that every variety has a limited range of adaptation. This range can be geographically impressive, such as the use of the same potato varieties in North Africa, Eastern Europe and highland tropical areas, but in other cases, varieties are confined to very geographically limited ecological niches. Apart from naturally occurring ecological conditions farmer-influenced conditions play an important role. In high-input agriculture, various ecological stresses can be overruled (e.g. irrigation, fertiliser application), thus adapting the ecology to the needs of the crop and its particular varieties. This results in 'wide' geographic adaptation, which is a very different concept as wide ecological adaptation. In low-input agriculture in marginal ecologies, the varieties have to adapt to the prevailing conditions. This illustrates the need for large numbers of location specific varieties in geographic regions of limited size. An important concept in this respect is agrodiversity (Almekinders et al, 1995). This concept combines ecological diversity, diversity in farm management aspects, and genetic diversity, including their interactions at different system levels. In the context of this study this means that genetic diversity (between and/or within) varieties in a particular farming system depends on the level of diversity in ecology and farming practices within the farming system, and on higher and lower level system levels.

> A good understanding of the choices that farmers make with regard to varieties is very important both in seed system support and in emergency seed provision. Supporting or supplying the wrong variety is very counterproductive.

1.3 Seed security

Definition

Seed security can be defined as:

The sustained ability of all farmers to have sufficient quantities of the desired types of seed at the right time. This definition includes a number of important parameters:
- Quantity
- Quality (desired types)
- Timing

- Finance ('ability to have')
- Equity ('all farmers')
- Sustainability

This definition is farmer-centred and not based on issues like national food security or rural development. The political will to assist farmers to increase their seed security, however, may be based on considerations of national food security or equity. It is however remarkable that literature on national food security strategies generally does not elaborate on seed supply (Anon., 1992; Hjort af Ornaas, 1992; Malambo, 1988; Sahn, 1989).

Seed security has always been a basic and continuous concern for farmers throughout the world. It becomes an issue for a wider audience only when seed is in general short supply due to 'unforeseen circumstances' at such a scale that it threatens agricultural production. The most striking examples of such 'circumstances' are:

* Ecological disaster;
* War situations, often combined with massive displacement of farmers,
* Disruption of the supply of seed from the formal sector to seed-dependent farmers.

Next to these obvious examples causing large scale seed insecurity there may be less obvious cases of seed insecurity at a smaller scale: Sperling *et al.* (1993) call the poorest bean farmers in Rwanda "chronically seed insecure" because they normally have to eat their stocks of food grain and seed beans and have to rely on whatever beans are available in the market at the time of planting. Such poverty induced seed insecurity is also widely experienced when relatively poor farmers have sudden expenses due to for example a disease in the family. These farmers can be considered 'acutely seed insecure'.

In both these cases of chronic and acute seed insecurity, individual farmers are likely to be able to acquire something to use as seed. This may be poor quality grain even though seed of acceptable quality may be available, but the farmers no not have the capacity to acquire it. Their seed insecurity thus concentrates on seed quality aspects rather than available quantities. Seed security can be analysed at different system levels: household seed security, and community, national and regional seed security.

Sustainable seed systems

The basic requirement for seed security in most conditions is the existence of sustainable seed systems. In most cases these are well functioning local seed systems, based on farmer selection and distribution within the community. National seed stocks and regional preparedness programmes are valuable and necessary, but basically only additional to a well functioning local seed system. In industrialised country agriculture, however, seed security is obtained through well functioning formal seed systems guaranteeing a regular supply of seed of acceptable quality to farmers. In many developing countries, however, public and emerging private seed suppliers are not always able to guarantee supply both in terms of quantity, quality and timeliness of supply. Seed insecurity due to disruption of the supply by the formal seed sector is a concern for hybrid seed and exotic vegetable seed users in particular, and more specifically for small scale farmers in remote areas using this type of seed. These are the first to be affected when available seed quantities are lower than usual. This can happen either as a result of poor domestic seed yields or of drastic changes in seed policy, such as sudden import restrictions. Seed security is generally best served by a diversified seed system, whereby the local and formal systems are well integrated. Current 'small seed enterprise development' efforts greatly support this diversification.

Emergency seed provision
Where the existing seed systems fail to supply sufficient quantities of seed, e.g. during or after war or ecological disaster, emergency seed provision can be part of relief operations. Quite often, seeds are only a small part of emergency relief programmes and the specific qualities of seed are often insufficiently acknowledged. The main issues are:

* When seed is brought in from outside the region, chances are that the genetic quality is poor, i.e. the variety may not be adapted to the growing conditions resulting in poor yields, or to the socio-economic conditions resulting in rejection of the produce. An example of the latter is the supply of yellow maize in Kenya, which was contrary to the locally used white maize not acceptable by many for human consumption.

* When seed is obtained locally, e.g. by purchasing the last stocks of food, the seed quality may be poor, resulting in poor germination, high levels of disease infection and weeds.

Seed security can be analysed at different system levels: household seed security, and community, national and regional seed security. Action to alleviate seed insecurity include seed system support and emergency seed provision.

1.4 Seed regulatory systems in developing countries

Where centralised seed production infrastructures were built to resemble the successful European and North American seed industry, external seed quality control institutions were developed to create a quality awareness with the seed producers and to safeguard the interests of farmers. The seed quality control institutions became the driving force behind the development of seed legislation. Their interest is based on the fact that a legal backing is needed when conflicts arise with seed producers or merchants.

In many countries these seed laws strongly resemble European regulatory systems, especially the English, French and Spanish in countries with colonial links with these countries. Such laws establish the institutional framework of national seed councils and certification agencies, and they regulate variety registration (release) and official seed certification. More recently, Plant Variety Protection (PVP) is being introduced in the regulatory system. Next to seed laws phytosanitary regulations and business laws may affect the seed system.

Seed council
Seed laws commonly regulate the tasks of a national seed board or council. The tasks are spelled out in rather general terms and basically include advising the Minister responsible for agriculture on all aspects relating to seeds, and overseeing the variety release and seed quality control systems. This offers a unique opportunity to arrange for all parties concerned to have a voice in the implementation of a seed law. Most national seed boards are however heavily dominated by representatives of the formal sector (in most cases even dominated by public servants). Research directors, heads of certification services and public seed enterprises may at best be complemented with a representative of a farmers' association and possibly with a representative of seed dealers or contract growers. The result commonly is slow and bureaucratic procedures for variety release and seed certification (see below) and little flexibility with regard to both emerging commercial seed producers and farmers that are best served by a diversified seed system. Having its roots in the formal seed sector, national seed boards are rarely directly involved in discussions on the conservation and use of agro-biodiversity. Being a national (!) seed board, seed security is commonly regarded at the national level, i.e. dealing with national security stocks and emergency provision of certified seed only.

Variety release
The variety release procedure generally has the following characteristics:

1. Application with a formal variety release committee and variety registration, including a variety description;
2. Testing for the value for cultivation and use (VCU) of the variety at different sites over a number of seasons, and possibly testing for distinctness, uniformity and stability (DUS); and
3. Analysis of testing results by the committee, followed by formal release.

Variety release systems are meant to assist breeders in testing their materials in different ecological zones, and to protect the national agricultural system from introducing varieties with poor agronomic characteristics and poor produce quality. Secondly, the variety release system provides valuable information for seed producers and seed quality control agencies on the important distinguishing characteristics of the new variety. Thirdly, the variety is linked to (commonly one single) name, which makes the market more transparent. A uniform naming of varieties can assist the identification of sources of seed in emergency situations.

Most variety release procedures allow for only a very small number of new varieties to enter the formal seed production system, and where certification requirements for uniformity are strict, the system also restricts the development and introduction of genetically heterogeneous varieties. There are, however, cases of acceptance of non-homogeneous varieties.

The Case of a multiline wheat variety in the Netherlands
The co-operative seed company 'Zelder' bred a multiline variety of wheat in the Netherlands in the early 1980s. A multiline is a set of near homogeneous lines, differing for one trait, e.g. a resistance gene. This novelty put a lot of pressure on the Dutch variety release system. The variety was insufficiently uniform, and could thus not be released. Finally, the option was to register the constituting lines as separate varieties. The resistance to yellow rust of individual lines was below standard, so they would not pass a VCU test. When tested as a multiline, however, the variety was superior. This meant that the system had to be flexibly interpreted. A proper participation of all parties, resulted in a final acceptance of the variety. Commercially however it turned to be a flop. The breeding and release procedure caused the multiline variety to be outdated before it even appeared on the market. Maintaining the lines and producing the seed separately were too expensive. The variety's name 'Tumult' (Dutch for 'uproar') was well chosen: it has caused a lot of commotion in the regulatory system. Zelder seed company has not produced multiline varieties since then.

Seed quality control
Official seed quality control is a useful addition to all seed procedures and methods used in formal seed production and conditioning that are geared to maximising the different seed quality parameters. The generation system (pre-basic, basis, certified seed classes), and prescribed isolation distances are used to safeguard the genetic seed quality: the varietal identity and uniformity have to conform to the varietal description. Procedures for seed production agronomy and conditioning are needed to optimise physiological (germination, vigour), sanitary (seed transmitted diseases), and analytical (purity) seed quality. Seed certification checks on such procedures and methods through inspections, seed testing, and labelling. The basis of field inspection is that a variety has to be identifiable and that off types can be recognised in the seed production field. This commonly leads to rather strict uniformity requirements for new varieties, and very strict variety maintenance procedures that keep the variety true to its original description. The reason is that it is very difficult to identify the number of off type plants in a heterogeneous variety. Seed testing is done to establish various seed quality parameters of the seed lot. This requires careful sampling of the lot and standardised testing procedures that are reproducible. There are basically two ways in which seed testing results are used. Some countries have seed standards, which results in a rejection of seed that does not meet the

standards. Other countries use the system of 'trueness to labelling', i.e. that the quality parameters are printed on the label and that the customer (farmer) can choose whether the quality is appropriate. Labelling is a basic requirement in a certification system: the label is used to identify each seed lot. It is used to control the varietal identity (the generation system) and to link the seed testing results with the total quantity of seed sampled. Furthermore it allows farmers to identify certified seed in the market.

Phytosanitary rules

Phytosanitary regulations are meant to avoid the introduction of new plant diseases and pests into a country (plant quarantine) and to avoid the spread of a disease in a country. Imported seeds and planting materials commonly have to be tested for the presence of diseases. This can be done either in the country of origin, which is then supported by a phytosanitary certificate, or imported plant materials have to be kept under quarantine conditions while it is tested for the presence of diseases and pests. Phytosanitary regulations may also prescribe a complete ban on the importation of particular seeds from prescribed countries.

Intellectual property and other rights over plant materials

In most countries, plant materials have been explicitly exempted from patent protection. The self-replicating aspect of plants is one of the reasons for the design of a special ('sui generis') system for protecting plant varieties. Also the fact that varieties cannot be described the way an industrial invention can be documented was the basis for what became the Plant Breeder's Rights system, later called Plant Variety Protection (PVP). The third aspect is that a variety can be used for the development of a totally different variety. These biological aspects and existing (uncodified law) tradition of the free use of plant materials, led to adapted criteria (distinctness, uniformity, stability), a farmers' privilege (free multiplication for own and non-commercial use), and a breeder's exemption (any variety is free for further breeding) have been introduced. Intellectual property rights have however not entered the field of plant varieties in most countries that are likely to have seed security problems on a regular basis. Due to the TRIPs Agreement of WTO there however is a trend towards the spread of both Plant Breeder's Rights and patent rights on either varieties or on biotechnological inventions.

PVP means that the breeder or discoverer of a new variety receives the exclusive right over the commercialisation of the variety. This means that the holder of the right can exclude others from the production and marketing of propagating materials. He can also licence other producers, commonly against the payment of a royalty. PVP only deals with new varieties; these are varieties that have not been commercialised before a particular date (relative to the date of application for the rights) and are valid for a prescribed maximum period.

Patents on biotechnological inventions have entered the field of plant breeding in recent years. Patents are now granted in various countries on both methods to modify genomes and on novel or transplanted genes. A debate is ongoing about the ethics of protecting genetic constructs, and especially on the ethics of protecting naturally occurring genes. Furthermore the discussion focuses on the scope of protection. Where plant varieties may be exempted from patent protection under the TRIPs agreement, patents on genes may 'de facto' result in patent protected varieties. Such varieties, even though they may be protected by Plant Variety Protection, may not be freely available for further breeding anymore, and also the farmers' privilege may not be valid. This is a cause for considerable concern.

The International Undertaking on Plant Genetic Resources for Agriculture (IU) has introduced the concept of Farmers' Rights, and the Convention on Biological Diversity the concept of national sovereignty over plant genetic resources. These concepts introduce some kind of ownership by farmers, communities or nations over 'old' varieties, including non-uniform landraces and other genetic resources. The operation of these rights has not been fully agreed upon as yet (June 1998). Various options are possible. On one end of the spectrum is kind of intellectual property over particular genetic resources (which thus have to be identified) of which breeders can make use after 'prior informed consent' with the community or nation of origin. This consent can be given, for example in return for a share in the profit on the commercialisation of the end product (the variety or the medicine/chemical compound). This option resembles a conventional Intellectual Property Rights system. At the other end is the pooling of all plant materials in one global system in which governments or seed producers have access to genetic resources under the condition that a global fund is supplied with sufficient amounts of money to safeguard the (in situ) conservation of plant genetic resources.

Additional rules on the international movement of seed

Various countries have special regulations in their national laws that affect seed trade. Quite regularly, the exportation of seed is prohibited when the national demand for seed has not been met. Export permits have to be obtained from a special office of the Ministry of Agriculture next to standard business permits related with exports and currency control. Zimbabwe introduced a regulation that exported seed has to carry an International Orange Certificate. This certificate can only be issued by the official seed laboratory, which may take a long time. Other countries even prescribe that exported seed has to be certified, i.e. that only formally produced seed of nationally released varieties can be exported.

In many countries, either the Minister or the National Seed Council is allowed to reduce seed standards or any other regulation in time of emergency. This aspect is very important for large-scale emergency seed provision, but does not apply to necessary movements of seed on a smaller scale.

Seed supply is commonly regulated by seed and phytosanitary laws, intellectual property rights and business laws. This framework is designed for the regulation of formal seed supply. Knowledge of these regulations is necessary in order to avoid frustrations in seed security support activities.

2. EFFECTS OF VARIETY RELEASE REGULATIONS ON SEED SECURITY

2.1 Variety release and diverse seed systems

It was noted above that variety release systems commonly restrict the number of varieties that enter the formal seed production system. There are also doubts about the value of the system for the correct selection of the optimal varieties for all farming systems in the country through such centralised testing.

Problems with the management of variety testing can be summarised under the headings Efficiency, Standards, Participation and Transparency (Tripp *et al*, 1996). Lack of efficiency includes both inappropriate site selection, and poor trial management. Poor trial management conceals actual differences among varieties, and as a result delay release due to the need for prolonged testing. In addition, trial management practices rarely represent common farmer agronomic practices. High input levels are commonly used as a matter of policy. Virk at al. (1996) report that average trial yields in the official 1989/1990 sorghum variety trials in India were three times the farmers' average yields (approx. 2800 kg/ha in trials against 900 kg/ha in farmers' fields). It may be doubted whether the trial results have any value to most farmers. Breeding is therefore still expected to target benign agro-ecological conditions only. Standards relate to the methods of trial layout, data collection and analysis. Most trials contain small plots, often with considerable border effects due to wide paths and other trial design matters, and in any case mono-cropping, even when farmers plant the crop in combination with others. Many variety trial stations are run by junior staff. They obtain precise instructions how to observe the characteristics of the crops. In some cases yield is the only important characteristic. In the best of cases diseases are screened, but these normally do not go further than marking presence or absence of a particular disease in the field. Quite often, characteristics that are extremely important for farmers are not taken into account, such as aptitude to intercropping, shattering (e.g. soybean), lodging when harvesting is delayed (e.g. maize), cooking time of the produce (e.g. beans). Breeding thus tends to concentrate on yield only without giving credit to the diverse needs of farmers.

Analysis of the trials is often done very objectively, i.e. using standard statistical analysis methods. The main complaint to this analysis is that many trials are pooled in one calculation. The variety having the highest average (!) yield is considered the best. This may, however, not be the very best variety in any of the testing sites. Standard variety release procedures rarely accept a variety that is specifically adapted to particular conditions. This explains the bias towards breeding for wide adaptation. On-farm variety trials are becoming more and more popular with variety release systems. This very positive development however hardly ever contributes to releasing more adapted varieties because such on-farm trials are either completely researcher managed (and thus similar to station trials), or the results cannot be easily analysed statistically. Especially the non-numerical comments by farmers are difficult to include in reports. Problems related to participation and transparency relate to the closed system that formal variety release often is. Parallel (demonstration) trials by the extension service, NGO's or private seed companies are rarely taken into account. Farmers are rarely well represented in the Variety Release Committee, or in the scrutiny of varieties in trials, thus leading to the very conservative trial designs and management.

Official variety release systems are geared to supplying varieties for the formal seed production system, which in turn produce seed for a market, i.e. for farmers that can afford to purchase (certified) seed. Variety release systems are not geared to selecting better varieties for remote and resource poor farmers, who may be served best with varieties that perform relatively well under low soil fertility and hazardous climatic conditions. Such farmers commonly value yield security more than high yield in good years. This way, variety release systems do not support resource poor farmers. The release system however may even counteract official rural development and biodiversity conservation objectives by affecting the direction of plant breeding towards the development of high yielding uniform varieties for benign agro-ecological conditions only. Since breeders are commonly rewarded on the basis of the number of officially released varieties, (participatory) plant breeding for more marginal conditions is often not taken seriously.

> Strict variety release procedures reduce the number of varieties available to farmers, direct plant breeding to concentrate on high external input farming systems and do not contribute to the development of diversified seed systems.

2.2 Variety release and emergency seed provision

Variety release procedures recognise a limited number of uniform varieties that are particularly adapted to mono cropping in rather benign agro-ecological conditions. Farmers, especially those in more marginal conditions need a relatively wide choice of varieties and in most cases not the officially released varieties (see 1.2). Emergency situations, especially those due to ecological disaster, are more likely to occur in such marginal conditions. It is extremely important that seed of the right varieties is provided.

Most emergency assistance organisations do not have detailed knowledge about seed and their quality characteristics. Especially the variety aspect is a rather specialised one and since variety information is only available on the few released varieties, such organisations commonly purchase seed of these known varieties only. Where such varieties may not be suitable for marginal conditions, farmers may not be optimally assisted, or even worse, they may invest land and labour for a crop that is bound to fail. Seed of non-released varieties may not be available in sizeable quantities. Such seed may have to be purchased from individual farmers, which requires a very good understanding of ecological similarity between the seed purchasing area and the target area. Apart from the limited knowledge there may be a legal argument that forces emergency assistance organisations to procure seed of released varieties only. In many countries only seed of released varieties is allowed "in the market". Even where organisations may be sensitive to the needs of the farmers in a particular disaster struck area, there may be no way to purchase seed of locally preferred varieties legally.

In longer-term emergencies, relief organisations may like to produce seed of local varieties in a safe area such as a neighbouring country. If these countries apply their own rules strictly they cannot allow the bulking of such seed unless the varieties grown in the disaster struck area are formally released.

Even though variety release procedures may produce valuable information on the varietal reactions to different farming conditions, and will select varieties for benign farming conditions, restrictive release systems can severely limit the availability of seeds that would optimally support disaster struck farmers.

3. EFFECTS OF SEED QUALITY CONTROL REGULATIONS ON SEED SECURITY

3.1 Seed certification and diverse seed systems

Seed quality control produces very important information for farmers, such as germination capacity. Problems may arise however when certification is made compulsory by law. When all seed has to be certified the choice of seeds is limited to a small number of released and uniform varieties with the results described above. Also, the possibilities for farmers to legally obtain seeds from other sources than registered seed companies are reduced. A strict seed law prescribing "all seed in the market" to be certified in fact turns sale or exchange of seed among farmers illegal. This means that farmers or farmer groups are not allowed to specialise in seed production without requesting for field inspection, seed testing and using officially produced higher generation seed as a starting material. The consequences are that it is illegal to produce landrace seed for sale (because these varieties are not officially released and not sufficiently uniform for certification). It also means that the one or few official seed suppliers are protected from competition to quite a large extent. Small seed producers may very well be able to produce a acceptable quality at a lower price because of smaller transportation and over head costs.

In recent years several organisations have started to support small-scale seed production activities. The objectives vary from 'improving farm saved seed', 'stimulating the use of diverse landraces', to 'bottom-up commercialisation of seed production'. Such projects are commonly not challenged by a seed law in the initial stages. There is however cases whereby seed laws have been used to frustrate such development oriented activities. The NGO, ENDA-Zimbabwe for example, was not allowed to multiply open-pollinated maize seed for export to neighbouring Mozambique, because in Zimbabwe only hybrids were released and the seed could not be certified. Since it is quite difficult for an emerging local seed business to adhere to all certification procedures and rules form the very beginning, seed laws may be a serious barrier to new entrants to the seed business which could increase competition for the benefit of the customer (the farmer).

Seed certification can be useful for protecting economic farmers from purchasing poor quality seed. Compulsory seed certification can however negatively affect diverse seed systems.

3.2 Seed certification and emergency seed provision

Regulations stipulating that all seed has to be certified plays an important role in emergency seed provision. Governments often insist that emergency organisations purchase seed for their programmes from official sources. This means that certified seed of uniform varieties has to be transported over long distances and distributed to farmers that may commonly produce genetically heterogeneous crops. In many cases it would be much more effective to look for remaining stocks of seed (often small stocks of food grain), clean and test it for germination and re-distribute that in the same area.

Also the organisations in many cases prefer to purchase a uniform product in bulk, because the logistics of buying many small quantities from individuals (without official stamps and signatures on cash receipts) is too complicated and the accountability too difficult to monitor. This means that it is not always seed laws that force the distribution of un-adapted seed, but also the lack of emphasis laid on the seed component of an emergency operation.

Another setback may develop when seed of local varieties of a disaster-struck area are to be bulked in a neighbouring country. Since the seed may originate from a genebank (very small quantities), very good care has to be given to such multiplication. The formal seed producers may be best equiped to quickly multiply the seed and to maintain a reasonable quality, but since it is not supposed to propagate non-released (national!) varieties, they may not be in a position to assist.

Regulations may force relief organisations to procure certified seed for emergency seed distribution operations. This may not be in the best interest of the target farmers.

4. EFFECTS OF INTELLECTUAL PROPERTY RIGHTS (IPRs)

4.1 IPRs and diverse seed systems

Intellectual property rights are meant to stimulate commercial plant breeding. The rights conferred to the breeder may officially restrict the diffusion of varieties from farmer to farmer. As such Plant Variety Protection (PVP) restricts the free use of varieties by farmers, but the farmers' privilege and the breeder's exemption limit potential negative effects from a legal point of view. The effects of patents on biotechnological inventions may however be significantly stronger, especially where (patented) genes may be included in local varieties that are important to (near) subsistence farmers. Patent holders are however not likely to exercise their rights with such farmers but there is a potential risk.

PVP does not seriously restrict diverse seed systems as long as the farmer's privilege, relaxed uniformity standards and the breeder's exemption can be maintained. PVP does however support the commercialisation of plant breeding, possibly reducing public plant breeding for marginal areas and resource poor farmers.

4.2 IPRs on emergency seed provision

Intellectual property rights may cause certain problems when for example large quantities of grain have to be transformed into seed in emergency situations. The holder of the right may ask for the payment of royalties, and he may even have the right to stop it especially when seed is 'produced' in one country to relieve another. The reason for a breeder to do this may be to protect his licence holder, who may sell less certified seed as a result of the large-scale use of grain. More likely however, the breeder will allow the use of grain, but only by his licence holder. This means that the whole procurement operation may have to go through this person who may misuse his monopoly.

PVP laws do however normally have an emergency clause in which a compulsory licence can be ordered. In all other situations, i.e. when seed of local non-protected varieties is purchased for distributed, IPR laws do not interfere.

5. DEALING WITH SEED REGULATIONS IN SEED SECURITY PROGRAMMES

5.1 Relief and development organisations

Seed support

Organisations that support farmers to increase their seed security through the strengthening of local seed systems operate within the sphere of the national seed regulatory framework. There have been cases that organisations have been hindered, such as through the rule in Zimbabwe that open pollinated maize varieties are not to be multiplied and sold. Less strong but still damaging is the situation in several countries that 'informal' seed programmes do not have access to the seed technology knowledge and facilities of the formal system, such as well-equipped seed laboratories. It is therefore important for NGO's and farmer groups to know the existing laws relating to seeds in a country. If it is not possible to act accordingly one has the option to just go ahead under the assumption that the legislator will not interfere as long as the local activities do not interfere with the formal seed markets. This is commonly so, but especially when an initiative becomes successful, opposition may grow.

Another alternative is to co-operate from the start with the formal institutions such as a national seed certification authority. Depending on a liberal interpretation of existing laws these can play a very positive role. An example is the role of the Seed Control and Certification Authority in Zambia, which assumed a role of supporter and co-ordinator of local seed system initiatives. Knowledge about seed production and technology are to be made available to informal groups as are the official seed testing facilities.

Emergency seed provision operations

Emergency production and distribution may be officially restricted. In disaster struck areas the legislator is, however, not likely to interfere either with or without an official emergency clause in the law. When not a whole country is affected by a disaster there may be strong pressure to purchase and distribute certified seed only. When seed has to be obtained in another country, problems may be even bigger. The exporting country may not have a good reason to interpret its own seed law pragmatically (because the disaster is not within its borders), thus forcing relief organisations to purchase certified seed. Secondly, even when non-certified seed can be obtained, the exporting country may still put seed quality and phytosanitary restrictions on the seed (which are more logically imposed by the importing country, but export restrictions are included in many national seed laws). Phytosanitary tests may seriously delay the purchase and distribution operations. Finally there may be business restrictions on seed exports; several countries have included a clause that restricts seed exports when the local demand is not fully met. So apart from the rather difficult issue of 'purchasing the good seed for the target group' there may be legal barriers to take. During a long term problems it is very useful to bulk local varieties from the affected area in a safe place (often another country) in order to distribute it at the time that farmers are likely to get back to their land. This was done within the Seeds of Hope programme to support war affected farmers in Rwanda. Such actions do generally require a good co-operation with the authorities, because strictly spoken, this may be illegal.

5.2. National Governments

Making seed laws more flexible

Seed laws definitely serve a purpose. It would be too easy to abandon the idea of regulating seed supply, given the obstructions noted above. Improving the situation regarding seed laws has to start with a proper representation in the National Seed Councils. These bodies could have a very positive effect in balancing flexibility and rigour in procedures and standards. A good basis would be to appoint at least 50% of the representatives of non-governmental organisations (farmers associations, private seed producers, dealers, importers, etc.).

Compulsory variety release systems could be transformed to variety information systems, whereby the emphasis is changed from the release decision to the testing of both new and local varieties. This means that variety testing is to be intensified wherever possible by co-ordinating all different sources of information: official VCU-tests, and also seed company, development organisation and farmers' experiences. An evaluation committee may extract statistical and non-statistical evidence of the performance of both new and existing varieties and landraces. The focus of the system has to be on publication of results, whereby farmers can choose for themselves how to use the information for their choice of varieties. The value of a variety testing system for seed security is illustrated by plans in Eritrea to systematically test materials from sources that could be useful in times of emergency. In this approach, the variety testing system is an important tool for disaster preparedness. Such activities may even contribute to the development of regional 'variety maps' which could be used in a Geographic Information system (GIS). Such GIS-linked variety informatio system could be very useful to identify seed sources in case of emergency. It could as a start be used to identify materials that could be included in disaster preparedness variety trials.

Seed certification systems have to be customer-oriented, i.e. those quality characteristics that really matter have to be prioritised, not those that are prescribed by a regulation. This basically means a voluntary system, whereby the facilities (inspection, seed testing) can be used by anyone producing seed. There has to be a tight control though on the use of official labels, and a strong information campaign about the meaning of the certification label. This way the certification agency has to prove its value. The uniformity requirement for certification may be relaxed through the wider use of the 'commercial seed label'. A draft seed law in Eritrea includes such a label reading 'produced from . . . (landrace name)' as an official designation. A draft seed law in Yemen contains a definition of the word 'seed' which excludes all farm produced and marketed seed from the rather strict regulations that are designed to guarantee the varietal identity and uniformity of formally produced and labelled seed. The bottom line is that not more should be regulated than what is strictly necessary. Both these solutions are within the boundaries of the OECD Seed Schemes that have been designed for the promotion of international seed trade.

An important support from a seed quality control agency to seed security initiatives is putting the facilities for seed testing and the specialised seed technological knowledge of its staff at the disposal of organisations and communities that want to improve their seed security situation (seed banking, local seed support initiatives). The Zambian plan to make the Seed Certification and Control Services a service point for all seed initiatives is highly commendable in this respect.

Finally, most countries have included an 'escape clause', in which the Minister may alter standards and allow other procedures in case of emergency. A survey in the West Asia, North Africa region (Louwaars, 1995) made clear that in virtually all countries such a clause exists, but that it is hardly ever used, despite regular emergency situations. In order to be effective, the Minister may be able and willing to act quickly when needed and not be bound to the prior advice from the National Seed Council, which may meet only once or twice a year. It is better even to leave enough room for emergency actions in the mainframe of the law to avoid undue delay of decisions.

Adapting intellectual property laws

Intellectual property rights do serve a purpose in the promotion of investments in private R&D. IPRs are a contract between society and the inventor. Common practice in the patent system is, however, more and more a struggle between competitors rather than a reward for an invention. IPRs have to assist in voiding monopolies that may distort free trade. Furthermore, IPRs in the field of plant varieties have to safeguard the non-codified rights of farmers to multiply seed of any variety for their own use and in the case of primary food crops for local diffusion. It may be understood that for export crops, countries are likely to win by granting a comprehensive type of IPR (e.g. flowers). This does mean however, that private investment in plant breeding of these crops will be very limited and that a publicly financed plant breeding will have to be maintained in many countries.

Protection of landraces through a PVP or other patent-like system does not seem useful. Landraces have been developed by generations of farmers/selectors, an it would not be sensible to protect these for a mere 20 or 25 years (as in PVP). Also, because of the necessity for protected subject matter to be identified, it would have to 'freeze' a landrace into a description. It seems more sensible to offer prizes to farmers or farmer groups who develop an "improved landrace" by further adapting it to the farming system. In the design of implementation rules of the IU and CBD, one should be extremely careful not to restrict the availability of genetic resources for both formal and informal crop development.

Making phytosanitary laws effective

Phytosanitary laws are important to avoid the importation of diseases and pests. As a result they have to be strict and implemented with optimum care. Especially in emergency situations, there is a risk of causing long term problems by introducing harmful organisms in order to quickly resolve short term ones. Phytosanitary regulations may however introduce additional bureaucracy, leading to delays or unnecessary destruction of seeds and planting materials. A further international harmonisation and a close contact between quarantine institutions will facilitate the acceptance of materials based on phytosanitary approval in the exporting country.

6. REFERENCES

Almekinders, C.J.M., N.P. Louwaars & G. H. de Bruijn. 1994. Local seed systems and their importance for an improved seed supply in developing countries. Euphytica 78:207-216.

Louwaars, N.P., 1994. Integrated Seed Supply, a flexible approach. In: J. Hanson (Ed), Proceedings of the ILCA/ICARDA Research Planning Workshop held at ILCA, Addis Ababa, Ethiopia, 13-15 June 1994, pp 39-46

Louwaars, N.P., 1995. Regional Co-operation in Seed Security in the WANA region. ICARDA, Aleppo, Syria

Louwaars, N.P. and G.A.M. van Marrewijk, (1996). Seed Supply Systems in Developing Countries. Wageningen: CTA.

Thomson, J.R., 1979. An introduction to seed technology. Leonard Hill, Glasgow, UK.

Wellving, A.H.A., 1984. Seed production handbook of Zambia. Department of Agriculture, Lusaka.

Tripp, R. 1997. New seeds and old laws. London: Intermediate Technology Publications.

Virk, D.S., A.J. Packwood & J.R. Witcombe, (1996). Varietal Testing and Popularisation and Research Linkages. Discussion papers series. Centre for Arid Zone Studies, Bangor, 27 p.

SEED STOCKS AND SEED MULTIPLICATION IN EMERGENCY SITUATIONS

Seed and Plant Genetic Resources Service
Plant Production and Protection Division
FAO, Rome, Italy

1. GENERAL CONSIDERATIONS

It is well known that good quality seeds of improved varieties, together with appropriate cultural practices, can make a substantial contribution to increasing agricultural productivity and sustainability at relatively little cost, if produced in adequate quantities, distributed in a timely manner, planted properly by farmers and priced appropriately.

The access of farmers' communities to seeds and planting material depends on formal and/or informal seed supply systems. The former generally consists of public sector research institutions, public and private sector seed production and marketing agencies, and seed certification and quality control organizations. The latter consists of large numbers of farmers who produce both traditional and improved seed varieties, market their production, and also, do their own research and development. For both systems, three factors have been shown to be critical characteristics of an effective seed supply system:

- Seed Availability - good quality seeds and/or planting materials must be available;
- Seed Delivery - transport and infrastructure must exist or be created to deliver seed and planting materials at the household, community, national and regional levels; and
- Seed Accessibility - existence of sufficient resources to acquire and multiply seed. Many farmers and communities are vulnerable to seed insecurity because they cannot afford or have no access to good quality seed.

2. SEED SUPPLY IN NORMAL SITUATION

2.1. Maintenance of Seed Stocks

In normal situations, plant breeding institutions and seed companies that have an important plant breeding programme maintain the basic breeding material and germplasm in different kind of collections: " Basic Collection" (long term conservation), "Active Collection" (medium term conservation) and "Working Collection" for daily use in the breeding programme. In addition, seed companies maintain a certain seed stock, the importance of which depends on the category of seed, reproduction rate and seed volume. As a rule of thumb, it is generally accepted that the breeder seed stock is kept in quantity to meet the totality of needs. The stock of pre-basic and basic seed should cover 50 to 25% of the amount needed for normal reproduction. Concerning certified seed, it could be considered that 10 to 15% of seed requirement should be kept in stock. Nevertheless, seed companies and producers always keep so called "carry over stocks" of unsold certified seed, which exceeds their own needs, and could be used in emergency situations. It should be mentioned that the maintenance of important seed stocks is very costly and risky, and unless Governments do not find its strategic interest to help, private companies are not interested in keeping large quantities of seeds in stock.

2.2. Seed and Planting Material Multiplication

The supply of good quality seed requires a well established seed industry based on modern and efficient seed technology, which includes all procedures and techniques applied in seed production and distribution, such as:

- Variety development and evaluation;
- Seed production, maintenance and multiplication of improved varieties;
- Seed processing and storage;
- Seed quality control and certification; and
- Seed legislation.

This is the usual procedure in providing good quality seed of improved varieties to farming communities and is known as the formal seed supply system or formal seed sector.

It has been estimated that although there have been substantial investments during the last 20 years or so in the development of the formal seed sector in developing countries, only approximately 5 to 10% of farmers' seed requirements are derived from formal seed supply systems. The rest is supplied by the informal seed sector, mainly through on-farm saved seed and exchange in farming communities.

Many farmers who rely completely or largely on on-farm produced seed are in areas where the sustained supply of improved seed enjoyed elsewhere does not reach. This is either because of geographical isolation or lack of commercial incentives for the private sector to reach such markets. This lack of incentives may result from either the negligible profit attached to the seed of some crop species, or high cost of distribution to more remote areas where organized marketing by the private sector is considered unrealistic.

2.3. Seed Distribution and Marketing

The establishment of an economically sound seed industry depends in a great extent on the market for which the seed is produced. The frequency at which the farmers will buy their seeds from the formal seed supply system will depend on different factors, such as:

a) the level of economical development of the country;

The Seed Replacement Rate (SRR) is the percentage of seed demand covered by the formal seed suplly sector. In developed countries, the SSR is much higher than in the developing or least developed ones. The formal seed supply system in developing countries is not sufficiently developed to make available good quality seed of improved varieties to farmers communities. The farmers in these countries, often cannot afford to buy seed every cropping season due to the low income from their farms, and are obliged to use their own on-farm produced seed.

b) the type of variety and the system of reproduction;

The seed of hybrid varieties of maize (single, double and three way crosses) must be renewed every cropping season to avoid the loss of hybrid vigour. The seed of improved open-pollinated varieties of cross pollinated crops must be renewed more frequently than the seed of pure line varieties of self-pollinated crops, if properly grown.

c) the commercial value of the crop;

The crops could be divided into three main groups, according to their interest for marketing:

- Cash crops such as cotton, cacao, coffee, tea, sugar cane, etc. The production of these crops is, in general, organized by state, parastatal or private companies or societies, including variety improvement and maintenance, seed production and distribution, supply of inputs and commercialization of the products.
- Crops of some interest for the market and agro-industry, such as maize, rice, soybean, sunflower, groundnut, vegetables etc. The government or public sector supports the research and variety improvement, and seed production and distribution, which sometimes is subsidized, making the private sector's involvement difficult. Nevertheless, national and international companies succeeded to produce and market hybrid and improved open-pollinated varieties of these crops.
- Staple or subsistence crops, such as pearl millet, sorghum, cowpea, etc, due to lack of a well organized marketing system, the crop is either self consumed or exchanged in the local market. The seed of improved varieties of these crops (hybrid varieties of sorghum and pearl millet are available in some developing counties) in many cases, is not purchased from the formal seed supply system and the farmers are using their own seed.

A well established seed industry must have an efficient marketing and distribution system, in order to supply seed and planting material to farmers in adequate quantity and quality, in time and at an affordable price. Private seed producers and dealers should have easy access to favourable financial resources (credits), and in order to be competitive, governments should not subsidize the public seed sector.

3. SEED SUPPLY IN EMERGENCY AND DISASTER SITUATIONS

It is important to recognize here that there are differences between seed sources which have seed stocks of varieties of major food crops available in limited quantities and seed sources which can supply adequate quantities of seeds without further need for multiplication before distribution to affected agricultural areas. Most often these seed stocks are for varieties which have come out of varietal improvement programmes and, although developed for adaptation to a wide range of agro-ecological growing conditions, may not be suitable for the replacement of local varieties due to a wide-range of farmer preferences that are not necessarily based on high yields. These different sources are discussed below.

3.1. Sources of Seed for Multiplication Before, During and After a Disaster

3.1.1. National and/or International Agricultural Research Centers

Both national agricultural research systems (NARS) and international agricultural research centers (IARCs) have crop improvement programmes which are aimed at developing improved varieties of major food crops adapted to a wide range of agro-ecological conditions. These improved varieties may or may not have undergone extensive on-farm testing to assure that these varieties are acceptable to limited resource farmers. In most cases, the quantities of seed available for any given variety will be limited depending on the extent to which they have been diffused for farmer-based trials and use by national extension systems.

It should also be stated that NARS and IARCs have neither a mandate nor the resources to multiply and store large quantities of seeds of adapted varieties. The mandate of these research systems is to develop and test varieties and when any given variety is well accepted, the seeds are normally then provided to other actors in the public seed sector to multiply and distribute, usually for sale.

3.1.2. Neighbouring Countries or Regions

Experiences from past emergency seed relief programmes have shown that it is preferable to procure seeds of local varieties in neighboring countries or regions which have seeds stocks available of local varieties which are adapted to similar agro-ecological growing conditions. Certain characteristics of these varieties are critical to assure that the seeds, when provided to affected farming communities, provide acceptable yields and do not waste the limited resources of these affected communities (time and labor). These characteristics include length of time required from sowing to physiological maturity, adaptation to altitude where grown and types of soils where they grow the best.

Often seed stocks are procurred from neighboring countries or regions through grain purchases on the local markets. Grain is not in and of itself necessarily good quality seed and should not be considered such. Large quantities of grain, when intended for seed relief operations, should be properly cleaned and tested for quality control (esp. for germination rates and insect infestation), and carefully marked for the important growth characteristics (if know) before distribution to affected rural communities.

3.1.3. National and International Genebanks and Collections

Genebanks and other germplasm collections are considered sources of seeds which have been collected, characterized and stored for use by public sector plant breeding and private sector plant improvement programmes. These are not generally considered as sources of seeds for seed relief programmes as the quantities of seed kept are quite small and would require extensive multiplication efforts to supply sufficient quantities of seeds needed to restore agricultural systems affected by disaster.

3.1.4. National and International Seed Producers (Public and Private)

National and international seed companies are the best sources of seeds for seed relief programmes as they have the capacity to either store adequate quantities or have the ability to multiply seed stocks of proven varieties of food crops. One of the main problems which consistently arises though is that public and private seed companies usually only deal with seed supplies of commercial crops which are cost-effective to produce and distribute, such as hybrid maize. They often do not deal with varieties of low value food crops which are usually needed to restore limited resource farming systems which use local varieties of cereal (sorghum and millet), legume (groundnut and cowpea) and root and tuber crops (cassava, yams and sweet potatoes).

3.2. Seed Supply During and After a Disaster

3.2.1. Emergency seed supply

Calamities and disasters of all kind cause interruption of agricultural production, and often result in the loss of seed and planting material of local varieties. In these cases, seed of adapted varieties is the major input requested by the agricultural sector to restore crop production systems. Usually, it is difficult to provide local communities with seed of such varieties, when they are more or less lost. Distribution of grain to be used for emergency seed supply produces low yields and increases the risk of introducing pests and diseases. Even if good quality seed of improved varieties, which are not adapted to local conditions is supplied, there is a possibility of low yields, or of non-acceptance by the farmers' community because of food quality or taste, which may not suit their requirements. In addition, the introduction of such foreign varieties induces uncontrolled crossings with local genotypes adapted to the region. This "genetic pollution" is responsible for the definitive loss of local genetic resources and, as a result, of the disruption of the sustainable agricultural system of the region concerned.

In order to alleviate the situation of lack of seed supply, which in several calamity affected areas often became critical, FAO with support of the international community, assisted the affected countries with emergency seed supply. According to FAO/TCOR data, approximately US$ 30 million was spent in 1995 and 1996 for emergency seed supply programmes (including seed purchase, transport, distribution and technical assistance, as well as, accompanying inputs such as fertilizers, chemicals and tools).

When planning for emergency seed supply, the following should be considered before deciding on the appropriate source of emergency seed: the varieties that are selected, the quantity of seed required, the quality of seed, the time requirements of the distribution programme and the price of seed from alternative sources. Once the crops for emergency seed supply are identified, attention must be paid to choice of variety. The objective is to return the local farming system to a situation as close to its pre-disaster status as possible. Choosing the appropriate variety(ies), if the local varieties are lost, is crucial for achieving this goal, and errors at this point may be responsible for exacerbating the emergency rather than ameliorating it. An updated database system with information of varieties suitable for the given agro-ecological zone and seed sources in the region is very important to prevent introduction of non-adapted varieties.

3.2.2. Restoring basic seed stock from local sources

Local markets could be a very valuable and reliable source of seeds of local varieties not only in normal times, but also in times of disruption due to disasters. For example, in Congo DR (former Zaire) after the looting during the unrest in the country in 1991, when the collection of local and improved varieties of the SENASEM were destroyed, the follow-up seed project managed to purchase grain stocks of local varieties in the rural markets and was able to restore a great number of lost varieties by using appropriate selection methodology.

In some cases, it is useful to identify grain stocks (public or private) of varieties of known origin and keep them separated from other grain as long as possible. In the case of emergency need for seed of these varieties, this grain could be cleaned, treated and used as seed. In Gambia, there was an emergency supply of groundnut seeds purchased from the local grain market in the framework of a TCP project. Before purchasing, the seed was tested for quality, and those seed lots which met the quality standards were purchased.

In some countries where a formal seed supply system is in place, comprising public and private seed companies and commercial enterprises, the companies could be approached and requested to supply seeds for emergency supply. Usually, many companies keep a certain quantity of seed, so called "carry over stocks", which exceeds in some cases their own needs, and could be used in case of emergency.

3.2.3. Purchasing from neighbouring countries and regions

In countries where local seed sources do not exist after a disaster, the seed should be purchased abroad. In such a case, the seed should be bought from suppliers from neighbouring countries, to save time and transport costs. If that is not possible, a basic principle is to buy seed from countries at the same latitudes, having similar climatic and ecological conditions, so that the varieties will be better adapted. It should be preferred to purchase seeds from legitimate seed companies rather than from grain merchants. Imported seed is also subject to possible delays because of transport problems, legislation (phytosanitary, plant quarantine and seed certification related to international trade) constraints.

When receiving donated seed, it must be kept in mind the origin of seed and variety. The varieties coming from the Northern countries may not be compatible with the needs of recipient countries in the South. Sometimes seed quality may also be low, due to a long period of storage before delivery.

3.2.4. Seed multiplication under contract

An alternative to buying seed is to contract for seed production. This requires considerable time, including arrangements of the contract, seed delivery and production itself, then collection of seed produced and distribution. In areas with possibility of irrigation or multiple growing seasons in a year, contracting seed production for the next season may be a viable solution. Where possible, tenders should be for supply of a given quantity of seed, as this can reduce costs significantly, and holds suppliers to a specified quality, date of delivery, etc. Seed production should be contracted to experienced seed producers, government seed farms or private companies, having adequate skills in seed production technology, processing and storage facilities that will ensure acceptable seed quality. When seed is contracted, it is advantageous to locate the production as close as possible to the zone of distribution. If seed production is contracted in a neighbouring country, it should be ensured that the legislation requirements concerning export and import of seed are fulfilled.

3.2.5. Multiplication of seed in neighbouring countries

In some countries affected by war for a considerable long time, a so called emergency across-border seed production could be organized. The case of Afghanistan is an example. In the framework of a UN/FAO project providing emergency relief and rehabilitation assistance, seed production for Afghanistan was organized in the neighbouring Pakistan, starting in 1988 for more than four years. After the situation in Afghanistan returned to near-normal, the project continued to rehabilitate the seed production system in the country.

3.2.6. Organization of national seed production and supply sector

In some countries where, during a long period of calamity situation (armed conflicts), the seed supply system was more or less interrupted or even destroyed, including the loss of human resources, seed production, processing, storage and marketing facilities, loss of genetic material and dislocation of people, once the situation returns to near-normal, there is a need to pass from emergency seed supply to the rehabilitation and development of the seed sector. In many countries recovering from a long period of a disaster situation, such as Afghanistan, Burundi, Rwanda, Liberia, Bosnia and Herzegovina, etc., the rehabilitation of the seed supply sector could not be achieved with the countries' own means, and international assistance must be sought. Donors are usually willing to finance such assistance, including purchase of material and equipment, restoration of seed production and processing facilities, training of national staff and launching seed production activities. In such activities, FAO plays a crucial role in providing technical assistance through all kinds of projects.

In such cases, various national and international agencies, NGOs etc., play an important role. Before the national seed industry and formal seed supply sector is rehabilitated and made operational, these agencies could be organized and used for launching the informal seed production programmes at the farmers' level (men and women). Training of extension agents, supply of quality seed of adapted varieties, small tools and equipment would be a good start for such an operation. When supplying seed, it should be kept in mind that if the local varieties and germplasm were lost, the first step should be to restore these varieties from local resources, or to introduce varieties from neighbouring countries, which are well adapted to the local agro-ecological conditions.

3.2.7. Strengthening institutional structures and capacity building

After a long lasting and devastating disaster situation, national institutions such as agricultural research, national seed service, seed testing laboratories, variety testing facilities, etc., are usually not functioning. The government administration and related services often do not exist, the public institutions, such as research, extension service, etc., cannot accomplish their tasks due to lack of qualified staff, adequate funding, equipment and other means necessary for normal work. Governments, in such cases, need external assistance and funding to rehabilitate and assist these institutions to return to their normal functions. In many countries, FAO has prepared national seed plans and programmes, and assisted governments in their implementation. In preparation of such programmes, it should be kept in mind that both formal and informal seed supply sectors should be included, with special emphasis to the role of private sector and on-farm seed production. In addition to the establishment and strengthening the national institutions, special attention should be given to education and training of seed staff at all levels, including seed growers and farmers.

3.3. Seed Information Systems

The basic information needed in order to organize an emergency relief operation is to know which crop varieties are most commonly grown in a given country or region which has been affected by disaster. There are many local varieties of any given food crop which have been developed by farmers over time. There are also improved varieties which have been developed by plant improvement programmes and introduced into rural farming systems over time. Information about these types of varieties varies greatly in its availability, especially inventories of local crop genetic diversity found in rural farming communties of developing countries, many of which are in disaster-prone regions.

3.3.1. Seed stock availability in the country and abroad

The FAO Information System on Crop Variety Enhancement and Utilization was started in 1969 from information gathered with the aim of assisting relief operations, evaluating and disseminating research results of plant breeding activities worldwide, and to promote crop diversification and utilization of potential useful plants. It is part of the World Information and Early Warning System on Plant Genetic Resources (see section 3.3.3 below on WIEWS).

The information in the Crop Variety Enhancement and Utilization System contains 51 groups of the 4,000 major crops of economic importance. Its compilation began as part of the FAO Information System on Seed and Plant Genetic Resources activities, and has the following main purposes:

- to assist locating suitable varieties to replace seed stocks reduced or lost in countries or regions affected by calamities;
- to improve variety evaluation activities in developing countries;
- to compile information so that, for each country, an inventory of the most commonly used and most useful varieties available can be prepared; as a result, an international data bank combining these national inventories was developed to promote the exchange of information about varieties;
- to promote seed trade between countries;
- to assist plant breeders, by providing updated information about varieties available on a worldwide basis; and
- to locate genetic material suited to particular purposes.

This rather unique database is continuously updated and is the result of gathering, analyzing and disseminating data on the enormous and ever-increasing amount of varieties available worldwide. As of July 1998, it contains about 170,000 entries but does not pretend to be complete; some vegetable varieties may be so common that possibly they could be available from hundreds of different sources.

A search of the database produces a list which is very simply structured that includes among other information: the crop variety name(s), the country(ies) from which the varieties are found, the importance of a variety in terms of area grown, and some indications are given of major traits like yield capacity, high protein content, average plant cycle from sowing to harvest, etc.

3.3.2. Seed suppliers

The FAO Crop Variety Enhancement and Utilization database goes together with the "World List of Seed Sources" (FAO Rome, 1996, 623 pages) which contains the names and addresses of over 8,200 public and private institutions dealing with seed production and processing in over 150 countries and includes a list of the major food crops with which the institutions are involved.

3.3.3. Availability of local germplasm in National and/or International Genebanks and Collections

The FAO World Information and Early Warning System (WIEWS) on Plant Genetic Resources system provides information on national programmes for the conservation and utilization of plant genetic resources for food and agriculture (PGRFA) and consists of a number of databases and information retrieval software including geographical presentation (maps) of location of *ex-situ* plant collections.

It provides a register of institutions involved in plant genetic resources conservation and utilization with a list of key scientists in each institution and a list of crops and activities dealt with. For each country, all institutions are given a series of numbers that hierarchically identifies them in the structure of the national plant genetic resources programme. The *ex-situ* collections database provides a summary record of each genebank's (collection's) holdings. These data include: species name and number of accessions for these species, type of material held (wild, land race, advanced cultivar, breeder's material, mutants, etc.), geographical distribution of accessions, and place of safety duplication. For each collection, the technical parameters of storage, such as temperature, humidity, moisture content and type of container are recorded.

The "Early Warning" database is still being developed and will consist of reports on the possible genetic erosion in the natural environment and *ex-situ* plant collections.

3.3.4. Other Early Warning Systems

Since 1975, the FAO Global Information and Early Warning System (GIEWS) on Food and Agriculture has been providing regular bulletins on food crop production and markets at the global level and situation reports on a regional and country-by-country basis.
The System aims to provide policy-makers and policy-analysts with the most up-to-date information available on all aspects of food supply and demand, warning of imminent food crises, so that timely interventions can be planned. To achieve this, the system:

- monitors food supply and demand in all countries of the world on a continuous basis. GIEWS' sub-regional analysts draw on twenty years of time-series data on food production, trade, food aid, stocks, and consumption. Current year data is constantly being up-dated as new information becomes available. Real-time satellite images are used as part of GIEWS' drought monitoring activities. Economic, political and agricultural information comes in from a variety of sources, ranging from official government releases, NGO evaluations and local and international media;

- compiles information and analysis of global production, stocks, trade and food aid and monitors export prices and conditions on the main grain exchanges;

- develops new approaches to early warning. Recent innovations include the development of an integrated computer workstation for cartographic analysis and a "Risk Mapping Project" to assist in the identification of vulnerable populations at the sub-national level.

- maintains close working relations with Governments, NGOs, other UN agencies, research institutions and the international media. The system depends on the free exchange of information and is supportive of efforts to improve information co-ordination;

- reacts to man-made or natural disasters by sending rapid food supply and demand evaluation missions to the affected countries and issuing once-off reports, which are quickly transmitted to the international community. The missions are often fielded jointly with WFP. The system has just completed a major revision of its methodology for rapid food security evaluations in countries where local data collection is non-existent or unreliable;

- reports to the international community on a continuous basis through its regular publications. "Food Outlook" and "Foodcrops and Shortages", have a global perspective and are available in Arabic, Chinese, English, French and Spanish. The quarterly report "Food Supply Situation and Crop Prospects in Sub-Saharan Africa" provides thematic, sub-regional and country-by-country analysis, focusing on sub-regions and countries which face food emergencies; and
- answers specific requests for information, from Governments, NGOs, research institutions and individuals.

3.4. Establishment and Maintenance of Seed Stocks

As mentioned in section 2.1., seed stocks maintained in normal situations depend on the category of seed, the reproduction rate and the volume. In any case, seed stock maintenance is very expensive and sometimes also risky. Therefore, it is not recommended to keep large quantities of certified seed in stock, since this requires additional costs, immobilizes funds for a considerable period, and in case of long term storage - the seed stocks could lose the germinability, and therefore, result in a sizeable economic loss for the company. Therefore, seed stocks should be considered as a strategic issue, and the governments should intervene and support their establishment and maintenance.

High category seed stocks (breeder and pre-basic seed) should be kept by the breeder or owner of the variety, if possible, under favourable conditions (cool rooms) in order to last for several growing seasons. The basic genetic material should be kept in long storage conditions (deep freezers).

3.5. A success story in Afghanistan

The main aim of the FAO seed programme from its onset in 1988 was to create a situation in rural Afghanistan conducive to the return of refugees and displaced families and start to redress the damage wrought by the war and assistance for agricultural rehabilitation. Food production and security are the essential and fundamental starting point and seed security and food security comes one after another in the sequence of an agricultural rehabilitation programme. The FAO seed programme sought to expand and sustain the use of improved seed through various activities- on-farm trials, selection, production, outside procurement, in-country seed production, networking, distribution, and popularization of certified seed. Fertilizer was also distributed to maximize the benefits offered by the improved seed. It was not very easy to implement the programme under an ever-changing security and political situation.

Such complex problems called for an imaginative implementation methodology of diversification and sustainability to remedy the situation rather than pursuing a typical programme. A vigorous and flexible calendar of operations for all stages had to be drawn up in advance and vigorously implemented. Even economic principle which dictates production of seed in a compact area was deliberately ignored to reach more farmers in isolated and inaccessible areas. The decentralization of the seed production was preferred. A diffuse approach consisting of large number of small seed producing pockets was adopted to ensure that farmers of every region and every stratum of society could benefit from the programme.

The FAO Programme has established in pragmatic ways that the provision of high quality seed and planting material (SPM) to Afghan farmers is one of the most cost-effective and economical forms of emergency aid. An integrated approach has been developed for marshalling, delivering and monitoring inputs. An informal back-up support network has been developed by the Programme. This encompasses the public and private sectors.

Collaboration with another UN agencies proved very fruitful. Food-for-seed was designed in collaboration of the WFP. The wheat seed was exchanged at a rate of 1:1.25 for wheat grain supplied by the World Food Programme. This ensured that seed growers had enough food to feed their own families and that a large proportion of their seed crops would travel out of the production areas to other farmers specially in food deficit areas. It also helped to make considerable savings of funds. A Fund Management Committee and a Seed Review Group were constituted to manage funds generated as sale proceeds and to advise on all matters relating to seed. More than US $ two million have been generated as sale proceeds and deposited in a separate account in Pakistan.

A number of constraints had to be overcome. Despite these constraints, more than 97% of the inputs did reach the farmers. By 1997, the Programme had distributed more than 48,413 MT seeds, mainly through NGOs and private voluntary organizations. The programme has been extremely successful and innovative. It continues to play a vital role in the rehabilitation of Afghanistan's agricultural production base after twenty years of conflict. As a result, it has been possible for many thousands of refugees and displaced families to return to their villages in Afghanistan with some expectation of being able to support themselves.

The FAO's Afghanistan seed programme provides many useful lessons. The most significant lesson of all that FAO has learned in Afghanistan is the importance of trusting the individual's own thrust for development. By keeping its intervention small and sharply focused on specific areas and problems it has provided the initial lift that the majority of Afghan farmers in need of before embarking on their own climb upwards to self-sufficiency.

4. CONCLUSIONS AND RECOMMENDATIONS

Recognition of the increasing cost and complexity of emergency operations has focused attention on taking preventative action and planning for disasters before they happen. When emergencies do occur, the emphasis is on moving as quickly as possible from relief to rehabilitation, reconstruction and sustainable recovery, so that affected households can free themselves from dependence on food relief, displaced populations can return to their own areas and refugee camps can be dismantled before they become permanent settlements. Now there is also a greater understanding of the need for close cooperation and partnership among the agencies involved to achieve a new, more integrated approach to humanitarian assistance.

Erosion of plant genetic resources for food and agriculture can occur in *ex-situ* collections, in farmers' fields and in nature. The former depends on the quality of the original material stored, and on the conditions under which the material is maintained and multiplied. The loss of wild relatives can occur through loss or disturbance of habitat or natural disasters. Loss of genetic resources in crops occurs mainly through adoption of new crops or varieties of crops with the consequent abandonment of traditional varieties without appropriate conservation measures. Various factors, both natural phenomena and the results of human behavior, including urban expansion, agricultural mechanization civil, strife and war can put plant genetic resources for food and agriulture at risk.

In order to minimize genetic erosion and its impact on agricultural production systems over the long-term, it will be increasingly necessary to monitor the key elements concerning genetic resources conservation and the various factors causing genetic erosion, and to assemble this information to enable remedial or preventive actions to be taken.

An early warning system for plant genetic resources is a necessary component of an effective seed security strategy as it monitors the underlying causes of genetic erosion at the national, regional and global levels. The challenge is to assemble the wide array of data and information that are available and to establish mechanisms to ensure that these data and information are transferred to appropriate points designated as responsible for analysis, coordination and action.

There are a number of Early Warning Systems which have been developed and implemented by FAO, USAID and other national organizations using a wide variety of data to provide an early warning of potential calamities, mainly due to unfavorable climatic conditions. An early warning system on plant genetic resources is a new approach to using these data to forecast potential disasters which could result in the loss of plant genetic resources within a given country or region.

There are a number of tools which have been developed and successfully used to predict crop growth responses to existing climatic conditions during any givern growing season. These include the use of FAO soils maps, overlaid using GIS with crop water requirements which when combined with daily rainfall data, can relatively accurately provide estimates of potential severe yield reductions and/or crop failures. Used as a proxy indicator, these data would provide an early warning for the potential threat of seed insecurity due to severe or total crop losses in disaster-affected regions.

This early warning would allow for the planning of mitigation efforts to prepare for and assure that sources of appropriate seed stocks adapted to these affected areas of a country or region are identified, procured, and multiplied for distribution to affected farming communities. This early warning would also help avoid significant losses of local crop genetic resources which may result from relief programmes providing inappropriate seed stocks in their efforts to assist in the re-establishment of crop production systems in the affected areas.

5. REFERENCES

Bay, A. 1995. A Seed Security Strategy for Africa. Paper prepared for AGPS. FAO, Rome, Italy.

FAO. 1996. Global Plan of Action for the Conservation and Sustainable Utilization of Plant Genetic Resources for Food and Agriculture. FAO, Rome, Italy

Fiebig, W. W. 1997. Developing Seed Security Strategies and Programmes for Food Security in Sub-Saharan Africa. Paper prepared for AGPS. FAO, Rome, Italy.

Lowars, N.P., and A.M. Marrewijk. 1996. Seed Supply Systems in Developing Countries. Wageningen: CTA.

ODI. 1996. Seed Provision During and After Emergencies. Overseas Development Institute, London, UK.

Scowcroft, W.R. 1996. The Role in Restoring Food Security after Disaster. Paper prepared for AGPS. FAO, Rome, Italy.

Sikora, I. 1997. Seed security - FAO Approach. Proceedings: Regional Technical Meeting on Promotion of Regional Network for On-Farm Seed Production and Seed Security in SADC Countries. Maseru, Lesotho.

Stevens, J.E. 1995. A Study on Seed Security Schemes. Paper prepared for AGPS. FAO, Rome, Italy.

FOOD AND SEED ASSISTANCE IN THE RECOVERY FROM CRISIS

Deborah Hines, Senior Policy Analyst, WFP
Sonali Wickrema, JPO, WFP
Leslie Van Straaten, Consultant, FAO

1. INTRODUCTION

Crises caused by conflict and natural phenomena profoundly affect the agricultural systems of rural people, destabilizing both their immediate food security and their ability to produce sufficient food for the future. Food and seed assistance in relief and recovery programmes can provide critical inputs for restarting production and for helping people to restore their livelihood systems. However, complementary programming is key if disaster-affected communities are to make optimal use of food and seed assistance.

When effectively programmed, the distribution of food, and then food and seeds, begins the process of strengthening coping strategies and moving towards recovery. The incorporation of development activities into relief and recovery interventions can help affected populations restore their traditional coping strategies sooner rather than later. Unfortunately, in many crisis situations, lack of continued access to land, insecurity or repeated natural crises impede crop production, making it necessary to continue to provide food and seed relief for a number of cropping cycles.

The effect of disasters on farmer seed systems was recognized by the 4th Food and Agriculture Organization (FAO) International Technical Conference for the Conservation and Sustainable Utilization of Plant Genetic Resources for Food and Agriculture, held in Leipzig, Germany in June 1996. A project entitled "*Develop institutional agreements and capacity to assist farmers in disaster situations to restore agricultural systems*" was subsequently launched, with funding from the Norwegian Government. From the first project-planning meeting, held in Rome, Italy, 2-4 February 1998 (FAO, 1998), a number of issues were identified as requiring further analysis. The issues relevant to this paper pertain to the relationship between food and seed resources, and their joint role in relief, recovery and development.

This paper considers the complementarity between food and seed interventions and presents an overview of the issues and measures needed for more effective joint programming. The paper begins with a review of the main stakeholders in recovery situations and their role in food and seed provisioning. Lessons about the operational aspects of joint programming are taken from both a literature review and case studies of three recovery situations - Burundi, southern Sudan and northern Uganda. The paper concludes with a discussion of the constraints to and opportunities for the joint programming of food and seeds.

For the purposes of this paper, the concepts of recovery follow those outlined in WFP's policy, *From Crisis to Recovery.* In a relief situation, WFP's mandate is to save lives. Following this immediate concern, through its food-aid-assisted recovery programmes WFP aims "to enable people to restore their livelihoods in order to assure immediate and longer-term food needs" (WFP, 1998a).

2. STAKEHOLDERS IN THE PROVISION OF FOOD AND SEEDS IN RECOVERY

2.1 WFP, Recovery and the Link between Relief and Development

Recovery policy and operational guidelines stress that traditional short-term relief responses are often inadequate in addressing the real needs of people who are trying to stabilize and secure their livelihoods in a crisis situation. WFP's approach emphasizes that:

- ◊ interventions need to move as quickly as possible from relief to rehabilitation, reconstruction and sustainable recovery;

- ◊ developmental approaches and activities must be introduced as early as possible into crisis interventions;

- ◊ crises resulting from natural disasters are often recurring, and longer-term approaches are therefore required, incorporating prevention, preparedness and early warning measures; and

- ◊ the transition to more development-based approaches is likely to be uneven, moving back and forth between relief/recovery and recovery/development, depending upon the security and stability of the area and the immediate needs of the affected population.

With a mandate for both emergency and development assistance, WFP is strategically placed to assist in the re-establishment of livelihood systems and the introduction of developmental approaches after a crisis. Through its three programme categories - emergency, protracted relief and recovery, and development - interventions start with immediate relief and move to more planned, longer-term investments in support of strengthening livelihood systems. WFP's provision of food assistance in conjunction with other inputs, such as seeds, also contributes to disaster preparedness and mitigation activities. The Vulnerability Analysis and Mapping (VAM) Unit provides information links with other early warning systems to help indicate when and where joint food programming is appropriate.

2.2 FAO's Emergency Response

FAO has identified eight distinct phases in the "disaster cycle", namely: prevention, preparedness, early warning, impact and needs assessment, relief, rehabilitation, reconstruction, and sustainable recovery. FAO plays an important role in each of these phases, as shown below. Specifically, it:

- ◊ helps prevent and prepare for food and agricultural emergencies

- ◊ provides early warning of their approach;

- ◊ assesses the impact of natural and man-made disasters on agricultural production and local food supplies;

- ◊ evaluates needs for immediate relief and longer-term recovery;

- ◊ delivers agricultural inputs and services to farmers in affected areas;

- ◊ helps countries plan and carry out agricultural rehabilitation and construction work; and

- ◊ helps countries design programmes and projects that will lead to sustainable recovery and prevent vulnerability to future disasters (FAO, 1997a).

Within FAO, the Special Relief Operations Service (TCOR) takes the lead in conducting assessments of agricultural relief and rehabilitation needs and organizing and delivering FAO emergency assistance. TCOR policy is currently being reviewed and updated. The Policy Assistance Division of FAO is responsible for the development of FAO's strategy in relation to agricultural rehabilitation, reconstruction and sustainable development.

2.3 Other Major Stakeholders in Food and Seeds Provisioning

The primary stakeholders in joint food and seed programmes are the affected communities. Their role in the provision of food and seed resources depends upon the scale of the crisis, the extent to which their social structures have been affected, and the capacity of implementing agencies to involve them. Where capacity exists, representative committees are formed through which communities help identify those most in need and distribute food assistance and complementary inputs. These committees are especially useful in voicing communities' needs for seed and other inputs to help them rebuild their livelihoods.

In most cases, the Government of the affected country is involved both at the national and local level in identifying the need for assistance and helping to coordinate the relief response. United Nations agencies provide support to governments and generally intervene within the areas of responsibility specified in their individual mandates. In most countries facing complex emergencies, the Office for the Coordination of Humanitarian Affairs (OCHA) coordinates the immediate relief response. The United Nations Development Programme (UNDP) or the United Nations Resident Coordinator helps coordinate assistance in situations moving from recovery to development. WFP is the principal provider of food assistance, and FAO's TCOR assists with the provision of agricultural inputs. The United Nations Children's Fund (UNICEF) plays a lead role in the provision of health support, water and sanitation, and supplementary feeding for children. The United Nations High Commissioner for Refugees (UNHCR) focuses on the protection and other needs of refugees and their resettlement and reintegration.

Implementing Agencies

The success of food and seed relief distribution depends on the effectiveness of the implementing agency:

- Food and/or seed distribution implemented by local Government structures have had mixed success. For example, the Zambian Government decided to rely on non-governmental organizations (NGOs), rather than on government agencies, to implement its "Programme to Prevent Malnutrition" in response to the 1991/92 drought. This decision was made for two reasons: to encourage donors to contribute generously to the programme; and to avoid the Programme from becoming "entangled" in the politics of the 1992 local government elections (Mukupo, 1994).
- Relief resources have been successfully distributed through community structures. In southern Sudan, Village Relief Committees have been involved in needs assessment and in the distribution of food aid provided by WFP with seeds provided by NGOs such as CARE (WFP, 1998c; USAID, 1996).

Donors provide support through their own programmes with the government, and by supporting the United Nations agencies and non-governmental organizations (NGOs). Donors either provide inputs (food and non-food items) directly or facilitate the purchase and distribution of relief inputs as well as medium- and long-term rehabilitation and development activities.

International and national NGOs usually act as implementing partners with United Nations organizations or bilateral donors. NGO involvement includes raising funds, providing resources, assisting with food or seed distribution, and designing and implementing programmes. Some organizations are involved solely in relief interventions, while others carry out both relief and development activities.

Information concerning suitable crops and varieties is provided by the in-country National Agricultural Research System (NARS), by seed companies, or by International Agricultural Research Centres (IARCs). National or regional seed companies, commercial suppliers, or local growers are the main sources of seed. Where FAO has a country presence, it helps implementing agencies identify appropriate seeds and seed suppliers and provides technical assistance on seed quality control and the timing of seed provision. If necessary, the IARCs may become involved in seed multiplication, where appropriate seed is not available commercially. Commercial suppliers are also providers of locally or regionally purchased food for food aid programmes.

3. FUNDING MECHANISMS

3.1 Donors' Funding Sources

Joint food and seed programmes tend to be financed through a mixture of funding mechanisms. Donors' funding mechanisms are attached to specific budget lines, which reduce their flexibility in providing joint resourcing. Often donors will have separate budget lines for food and seeds in relief and in development programmes.

The current donor climate makes it extremely difficult to maintain responsive programmes outside the parameters of relief. Dividing resources into "development" and "humanitarian" precludes the funding of longer-term strategies in areas classified as humanitarian. For example, donors may use humanitarian funds to support the provision of food and seeds after a crisis. However, if an agency wishes to develop seed multiplication sites to establish local sources of seed, donors may be unable to provide funds as this is considered a "developmental" activity.

Most donors fund rehabilitation or transition components within their relief programmes but are careful not to label them as development activities. Many of the relief activities and implementation strategies supported by donors are in fact transitional in nature and are rooted in development. For example in the Sudan, the UNDP Area Rehabilitation Scheme is a good illustration of a transplanted development approach that has been introduced on a pilot basis in villages affected by war and in areas where returnees are concentrated. Thus transition activities are being implemented, albeit under the following major constraints:

◊ short-term funding windows - generally of 12 months or less - which involve projects that:

- ♦ can demonstrate quick impact and are administratively and technically simple; and
- ♦ have capacity-building potential (participatory, credit, training).

3.2 Resource Mobilization

Donors are willing to support NGOs or United Nations agencies in emergency agricultural relief. In some cases, where seeds are being programmed for food aid recipients, donors have provided additional funds to NGOs or other United Nations agencies to pay for WFP to transport seeds and tools with food aid deliveries (OCHA, 1998). In recovery situations, WFP encourages partners to submit project requests for seeds to enable WFP to commence a phase-out strategy. Such requests clearly state that seeds and tools will be jointly programmed with WFP food.

In general, FAO can only finance the provision of seeds and technical advice through its regular programme channels. However, in an emergency, FAO is able to access limited funds for the provision of a coordinator to provide implementing agencies with technical advice and coordinate agricultural relief operations. These funds can only be used within the first year of an emergency intervention. Beyond this, FAO is reliant on direct donor support, especially in the case of protracted emergencies.

WFP can mobilize resources both in-country and through its headquarters. At the onset of a crisis, the Immediate Response Account (IRA) can be used to mobilize resources quickly to start an operation, but funds must be repaid once contributions from donors are received. Contributions include food commodities or cash for the purchase of food. In addition, donors are required to meet the full project support costs of programming each ton of food.

Non-food commodities can be included as a project support cost if they increase the effectiveness of the food intervention. Funding is usually limited to cooking utensils and kerosene, and in a limited number of circumstances, seeds and tools. WFP prefers to mobilize funds for food to be provided in support of seed programmes undertaken by governments or partner agencies.

The extent of coordination in resource mobilization and joint programming depends on mutual need. For example, UNHCR and WFP often coordinate food and seed programming as UNHCR may provide seeds within a food aid phase-out or reduction strategy. In such cases, WFP's future funding requirement for food is dependent on UNHCR's success in the timely and effective provision of seeds. In contrast, FAO mobilizes funds for seed programmes and, if needed, requests food for seed-protection rations from WFP. Where inter-agency communication is effective and local institutions are involved in programming, agencies are better able to coordinate the joint provision of food and seeds.

3.3 Consolidated Appeal Process

Food and seeds can be provided in response to emergencies caused by either conflict or natural disasters. At the onset of a major emergency, the Consolidated Appeal Process (CAP) provides a coordinating mechanism for the resourcing of food and seed programmes. Agencies request and receive funding mainly for the resources they distribute: food, seed, medicines. This allows transparency and accountability in the allocation of donor funds.

Consolidated Appeal Process

The United Nations Consolidated Appeal Process (CAP) is a formal mechanism for the joint planning of and appeal for emergency assistance. The CAP is overseen by the Office for the Coordination of Humanitarian Affairs (OCHA), with the formulation of the appeal grounded in a country-level coordination process. In terms of resource mobilization, the CAP provides donors with a single intervention plan for all United Nations agencies working in a country, but each component of the plan has a separate budget. In this way, individual United Nations agencies have their own funding requirements corresponding to their commitments in the intervention plan. As an example, food and seeds may be jointly programmed but WFP will request food resources and FAO will request funds for seeds.

The CAP allows donors to see how food and seed programming has been planned in relation to other United Nations activities in a given country. However, as funding is by component, one agency may receive funding for one component while another agency, with a complementary component, may receive insufficient or no funding. In most cases, WFP's relief food assistance receives funding. However, inputs such as seeds or training, which are provided by other agencies, are not always funded. A problem with CAPs has been that donors support budgets for the immediate humanitarian response, but exclude funds for the recovery portion of the intervention plan. This is one reason why food is generally more readily funded in CAPs than seed inputs.

The CAP is currently under review, with the aim of making it a more effective and strategic tool for relief coordination. The intent is to decentralize the process further and increase planning, implementation and monitoring functions at country level. It is still too early to assess the potential benefits of the proposed revised CAP process. However, there is optimism that improved country-level planning will improve the complementarity between activities, and the revised CAP will be used to compile needs assessment information and incorporate NGOs' activities and funding requirements in the intervention plan.

4. REVIEW OF JOINT FOOD AND SEED OPERATIONS

A literature review and three in-country case studies (in Burundi, the Sudan and Uganda) were undertaken to develop a better understanding of effective coordination mechanisms for joint food and seed programming. This section highlights experiences that are relevant to the successful implementation of food and seed relief interventions and discusses operational considerations for the joint programming of food. Given the scarcity of literature relating to the joint programming of seeds and food in a relief/recovery context, many of the lessons are extracted from the three case studies.

4.1 Literature Review

A list of actions necessary for effective joint programming of food and seed relief interventions is presented in the following box. These actions have been summarized from a number of sources and are presented as a series of steps. The sources include:

- The Overseas Development Institute (ODI, 1996), which describes the components of seed provision during and after emergencies, highlighting where food assistance may be necessary;

- A United States Agency for International Development (USAID) review of seed interventions in the Greater Horn of Africa, which provides recommendations for seed provisioning in the context of moving from food aid to food security approaches (USAID, 1996);

- Campbell and Harvey (1998), who present the problems experienced in southern Sudan as agencies attempt to move from relief food aid to more "developmental" interventions; and

- Pottier (1996), who examines the different modes of food and seed provisioning in Rwanda after the crisis of 1994.

To ensure that food and seed programmes make a positive contribution to the food security of target populations, it is essential that the activities supported match with the priorities of communities in affected areas. In addition, selected crops should represent priority food crops of the region and the varieties must be appropriate to the agro-ecological zones targeted (ODI, 1996; USAID, 1996). Inappropriate activities, crops or unadapted varieties result in crop failure, the demoralization of beneficiaries, serious delays in the planned withdrawal of food aid, and can have negative long-term consequences on the cropping system (Richards and Ruivenkamp, 1997).

Although food and seed distributions are often an essential step in stimulating local food production, long-term interventions can have serious detrimental effects on local agricultural production, on traditional seed supply systems and on the local economy. Such negative effects were experienced in Zimbabwe following the prolonged blanket distribution of food, fertilizer and seed during the early nineties (Friis-Hansen and Rohrbach, 1993) and in Somalia during 1993 and 1994 when the continued arrival of food aid prevented farmers from selling their agricultural produce (Grunewald, 1996).

Sequencing of Joint Food and Seed Relief Interventions

Important Steps include:

- evaluate the nature and extent of the crisis and review possible responses in a coordinated forum;
- determine the number of families affected, their needs and type of assistance required to meet these needs;
- determine if relief food and seed distributions are needed;
- determine for different target groups if it is more appropriate to provide food before/with/after the seed distribution or continuously until the following harvest;
- ensure that necessary, locally adapted, complementary inputs are provided (hand tools, non-food household items, health support, nutritional programmes);
- identify roles and responsibilities of implementing agencies for the provision of seed and food assistance;
- identify appropriate crops and varieties to be distributed and appropriate sources;
- give the beneficiaries prior warning about food and seed distribution, including information on the timing of deliveries and the commodities and crops/varieties to be distributed;
- explain that the distribution of seed is part of a strategy to reduce future food and seed distributions;
- implement the food and seed distributions according to the logistics plan;
- evaluate the effectiveness of the food and seed distribution, especially issues of timing; and
- monitor crop development and yields as a means of estimating future food requirements.

In southern Sudan, the Agency for Cooperation and Research in Development (ACORD, 1995) noted that during 1987/1988, food production in the town of Juba was depressed due to continued supplies of food aid. Learning from this experience Operation Lifeline Sudan began incorporating food security aspects such as developing seed systems to relief interventions.

After the priority of saving lives is met, a first step in the transition process for WFP is to better focus food aid by: improved targeting and definition of population groups and geographical areas in need; distinguishing between different sets of needs; and tailoring food rations to need, for example by providing food aid on a seasonal basis. In order to avoid dependency, WFP, like most organizations, strives to move from blanket food distributions to more targeted distributions, without neglecting the provision of assistance to the more vulnerable groups. Food for work (FFW) is the most common mechanism for this.

A number of seed-related activities can be introduced as the transition process moves away from the need for relief assistance. NGOs with agricultural technical capacity tend to move quickly from seed distribution to development-oriented activities. In the early eighties the development programme of ACORD in Gulu District, Uganda, moved from the free distribution of hand tools through the Western Acholi Cooperative Union to support of the Union's engineering workshop to help develop local capacity to produce agricultural equipment (ACORD, 1995). Seed multiplication activities also help to re-establish seed supply systems, while agricultural extension activities help to promote sustainable agricultural techniques and improved cultural practices (USAID, 1996; FAO, 1997b).

Variety Selection

During the late eighties, agencies implementing emergency seed distributions to war-affected Mozambican families were reliant on imported seed. Although the imported varieties were acceptable, they were not always the most appropriate for local conditions. In 1991, World Vision (WV) together with the NARS, Instituto Nacional de Investigação Agronómica, initiated an on-station and on-farm research programme in parallel to its emergency interventions, to identify better adapted varieties. A wide range of varieties produced by the NARS, the national seed company, regional plant breeding companies, and IARCs were evaluated. Varieties were selected on the basis of early maturity, pest and disease resistance, drought tolerance, adaptation to low fertility, taste/cooking quality, as well as yield. The short season maize variety Matuba was shown to be particularly well adapted to the needs of the internally displaced and drought-affected farming families in central and northern Mozambique (Sperling *et al.*, 1994). The longer-season variety Manica was shown to be appropriate to the higher rainfall upland areas (Sitch *et al.*, 1996). As a result of this research programme, WV provided information on appropriate varieties to many NGOs planning seed relief interventions.

Surveys concerning the fate of varieties disseminated by WV suggest that, as a result of WV's relief distributions and farmer-to-farmer seed exchange, more than 600,000 families in central and northern Mozambique are now growing Matuba.

Source: Chapman, White and Nankam (1997)

4.2 Case Studies

The case studies that form the basis of this paper are part of two separate efforts looking at issues of rehabilitation and recovery. The case studies in Burundi and Uganda looked specifically at joint food seed programming issues in situations of conflict and natural disaster. The studies were undertaken by WFP with funding from FAO. The Sudan case study was part of a larger WFP effort looking at programme implementation issues in rehabilitation and recovery, of which seed and food programming was an important component. While all three countries combine relief and rehabilitation activities, the contexts in which food and seed resources are provided vary considerably. The main findings from the three in-country case studies are discussed below, followed by a discussion and summary of the main lessons learned from the literature and case studies.

Burundi case study

The programming of food and seed resources is relatively successful in Burundi, principally because FAO's Emergency Coordination Unit has established a coordination mechanism that fosters effective inter-agency relationships. The important elements of sound coordination, including joint needs assessment and targeting, are discussed with respect to the joint programming of food and seed resources.

Emergency Response Coordination in Burundi

Context

Prior to the conflict that began in October 1993, Burundi was generally food secure. The 1993 crisis led to the massive internal displacement of local populations (more than 600,000 people are currently internally displaced (WFP, 1997)). Families living in regroupment camps and recently displaced populations have limited access to their land due to rebel activity, and require intermittent food assistance. Although the long-term internally displaced persons (IDPs) have developed some coping strategies such as working on tea plantations, producing beer or negotiating access to local land, many remain vulnerable to fluctuations in agricultural productivity and ethnic attack. The hilly terrain and high population density of Burundi mean that IDPs often have to farm rainfed marginal lands and often have to negotiate access to land on a seasonal basis.

The economic situation in Burundi degenerated after an embargo was imposed by neighbouring countries in June 1996. Market prices for food and agricultural inputs more than doubled as imports fell. Also, due to the volatile nature of conflict between factions, some provincial markets have disintegrated and marketed food and seed supplies have disappeared. IDPs face the dilemma that cash incomes are eroded by inflation and goods may not be available, while saved food and seed stocks may be raided by militants or have to be abandoned when fleeing attack.

Coordination Mechanisms and the Role of Different Stakeholders

The FAO Emergency Coordination Unit for Agricultural Relief Operations is responsible for the national coordination of the emergency response for agriculture. National coordination meetings are attended by the main United Nations and NGO implementing agencies as well as two government ministries - the Ministry of Agriculture and Research and the Ministry of Reinsertion, Reinstallation, and Regroupment. The meetings help build a consensus on provincial seed needs and allocations. Implementing agencies are encouraged to share resources to ensure a more complete coverage of the needs. A "lead agency", assigned to each province, is responsible for provincial level coordination.

WFP is the largest provider of food assistance. Fluctuations in food insecurity have prompted WFP to adopt a strategy of ad hoc interventions as well as supplementary feeding to vulnerable groups. Ad hoc food distributions and FFW programmes are carried out either by WFP or by implementing NGOs. WFP partners with both UNICEF and UNHCR to support vulnerable groups. In addition, WFP provides logistical support for seed distribution, which is funded by UNDP/United Nations Office for Project Services.

In collaboration with the national agricultural research system, ISABU, FAO provides guidelines concerning appropriate crops/varieties, seeds/tools quality specifications and recommended suppliers. FAO has also assumed responsibility for the quality testing of all seed purchased in Burundi. FAO monitors market prices to determine when and where bean seed (the principal emergency crop) should be purchased.

Needs Assessment and Targeting

Food assistance needs are determined by WFP through an adaptation of the Save the Children Fund (UK) Food Economy Approach (Boudreau, 1998; McDonagh, 1996). Information concerning the origin of children admitted to therapeutic and supplementary feeding centres helps WFP target the most food-insecure populations. Pre-harvest FAO/WFP crop and food supply assessment missions also help identify the areas most vulnerable to food insecurity and estimate future food aid needs.

Seed needs are coordinated through FAO. Seed distribution focuses on locally purchased bean variety mixtures and vegetables. Only food-insecure populations with access to at least half a hectare of land receive seeds. FAO liaises with WFP to determine where food rations need to be given with seeds.

Joint Programming of Food and Seed Resources

Joint programming of food and seed resources occurs in three ways:

- Seed-protection food rations are provided to seed beneficiaries living in food-insecure areas. Such families are identified jointly by WFP, FAO and their implementing partners.
- Demonstrations of vegetable production are given at supplementary feeding health centres and vegetable seeds are provided to families when they are discharged.
- Food for work is also being used to encourage women's groups to cultivate vegetables. Food rations are provided for two months, covering land preparation and vegetable cultivation, and thereby helping to build up the health of the participants and maximize their input into vegetable production.

Problems Encountered

- The needs assessment of WFP's Food Economy Approach is constrained by insecurity and lack of access to reliable informants.
- Logistical constraints caused by limited trucking capacity in-country, the narrow window of time in which to deliver seeds, and volatile insecurity have caused difficulties for the timely delivery of food with seeds and tools.
- The high turnover of NGO staff (particularly within the provincial "lead agency").
- Although WFP-Burundi would prefer to provide a 30-day or 15-day seed-protection food ration, food pipeline problems during the last season limited the proportion of seed beneficiaries receiving seed-protection rations and reduced the size of such rations from 15 to 7 days.
- The local purchase of large quantities of bean variety mixtures is hampered by supply limitations, transport, and insecurity problems. The embargo has also caused pipeline problems for FAO in obtaining vegetable seed and agricultural inputs in time for scheduled food deliveries.
- As beans are often stolen in insecure areas and there is a strong export market for them in Rwanda, many farmers sell their produce rather than consume it or save seed. FAO is, therefore, considering whether it would be more appropriate to distribute another protein-rich crop as an alternative to beans.

Sudan case study

Since 1956, the year of the Sudan's independence, the country has experienced only a decade of relative peace, between 1972 and 1983. Concurrent to the war, drought and ecological disasters plague the Sudan; the resulting famines of 1973-74, 1984-85 and 1998 stand out as major natural disasters. Operation Lifeline Sudan (OLS) was established in April 1989 as a United Nations coordinated programme developed in agreement with the Government of the Sudan and southern independence movements.

The joint programming of food and seed resources has faced political, financial, and coordination problems, to a large extent caused by the uncertainty of how to respond to a complex humanitarian situation. Thus, the move to recovery is not uniformly supported by all donors, the Government and the different movements. The Sudan case study provides an interesting example of coordination, planning and implementation in a war situation exacerbated by natural disasters.

Experiences From Aweil East County, Bahr-al-Ghazal

Context

The province is intermittently affected by conflict (militia activity, incursions from the north, marauding from trains, inter-factional fighting) and certain locations are controlled by warring parties. Villages are periodically looted and burned, cattle raided, and crops stolen and burned. Natural disasters (drought, flooding and environmental degradation due to restrictions on cattle movements) also periodically disrupt crop harvests. This county is inaccessible during the rainy season and villages can only be reached by footpath in the dry period owing to limited infrastructure. Because of insecurity, airstrips are not always operational.

Traditionally, Aweil East County was a surplus grain area. In recent years the area has experienced grain deficits and in 1996 there was a reduction of about 25-30 percent in harvest yields. In difficult times, communities are forced to expand their coping strategies and rely on collecting wild fruit, fishing and bartering livestock for grain (primarily sorghum or dura). Access to basic services and markets is poor and the area suffers from widespread food insecurity. Primary education and poor health facilities are available in most villages.

Coordination Mechanisms and the Role of Different Stakeholders

Through OLS, United Nations agencies and NGOs work in collaboration with the humanitarian agencies of the Government of the Sudan and the southern independence movements to ensure access and the delivery of humanitarian assistance to all war-affected people. The operation is divided into a northern sector, serving civilians in government-held areas, and a southern sector for areas held by the Sudan People's Liberation Movement (SPLM), the South Sudan Independence Movement and the SPLM-United. There are currently more than 40 international and Sudanese NGOs working under the umbrella of OLS. UNDP is the lead coordinating agency for the northern sector and UNICEF takes the lead in the southern sector. WFP has been providing emergency food assistance in both sectors through its country offices in Khartoum and Nairobi, while seed interventions are undertaken by UNICEF and NGOs. Where feasible, WFP delivers seeds for UNICEF.

Village Relief Committees play an important role in ensuring that beneficiaries are represented in the various aspects of programming: needs assessment and identification of inputs, targeting, distribution, and monitoring. These committees - composed of seven women and six men - play a crucial role in establishing links between the affected community members, the donor agencies and the various local authorities representing groups within the SPLM.

Security and logistical constraints limited WFP distributions in the country from January to April 1997; only 30 percent of planned deliveries were made and 42 percent of intended beneficiaries were reached. Insecurity limited the number of flights WFP was able to make and the locations in which it could operate. In addition to insecurity, flight capacity was a primary constraint during this period as WFP had only one Buffalo airplane for food distributions. These constraints forced WFP to reduce not only the number of items in the food basket (in order to deliver larger quantities of cereals), but also the ration size, to be able to reach more people with each food drop. In addition, support to seed programmes had to be cut as food was targeted to priority humanitarian needs.

Needs Assessment and Targeting

In the southern sector, WFP works with Save the Children Fund (UK) to assess household food economies. This gives a picture of how families cope with food insecurity and where food and seed provisioning may be appropriate. In the northern sector, WFP uses VAM in coordination with USAID's Famine Early Warning System (FEWS) to assess communities' needs. The Village Relief Committees provide the most detailed information on areas and beneficiaries requiring food assistance. The committees visited appeared to understand their role of targeting the most needy for food assistance and were aware that they were accountable for their decisions. They also were quick to point out the other needs of the community and the fact that they were trying to secure other inputs for their villages. The possibility of setbacks and reversal is an accepted factor of working in the area. The local people know that their survival depends on their ability to restore their means of production in the long run.

The committee process is considered more efficient than targeting through chiefs because: the share or tax taken for the Movement is less; the food is more equitably shared among those most in need; families now receive food through women, thus a husband does not have to distribute between all his wives - each wife receives according to her need; and there is less disruption by the civil authorities in the distribution process. Conditions impede follow-up monitoring to verify if and how much food or seeds were reaching intended beneficiaries or to determine the effect assistance had on household food security.

Joint Programming of Food and Seed Resources

Seeds programmes have, in general, been developed independently of food aid although joint provision is increasing as agencies attempt to increase the population's food security despite the difficult political situation and recurrent natural calamities (Macrae *et al.*, 1997). In the north, UNICEF, FAO, UNDP, and NGOs are all involved in seed and tool programmes. In the south, UNICEF and NGOs are the main providers of seeds and tools. Food and seeds are programmed together in relief interventions - especially where beneficiaries have few or no options for coping with food insecurity. However, donors are less willing to finance food and seed programmes that appear to be developmental. The line is fine and varies from agency to agency. In its Integrated Strategic Plan 1997-99, USAID allowed that funding seed research and the introduction of new crops was consistent with the humanitarian context of southern Sudan. The plan also supported the joint programming of food aid with seeds to ensure that food and seeds were delivered in a timely manner to enable target groups to build local capacities for food self-reliance.

Problems Encountered

- Food aid to the Sudan is financed principally from humanitarian budgets. Donors are politically unable or unwilling to finance the provision of seeds, and especially seed multiplication projects, as these are not considered humanitarian. This has led to the development of seed programmes separately from food aid assistance. Since 1996, an effort to move from food aid to food security approaches has been made by OLS with mixed support from donors.
- Limited logistic capacity of WFP and partners (aircraft, for example) result in pipeline delays and the inability to reach the beneficiaries or in the need to reschedule non-food items within the distribution plan. This situation often disrupts planned joint seed and food deliveries.
- Lack of coordination between seed and food distributions. The Village Relief Committees emphasized that food was not just a tiding over measure during times of hunger gaps. They stressed that food should be distributed just before the seeds, to preclude the eating of seeds and ensure that beneficiaries have the energy to undertake productive activities such as preparing land and planting the seeds.
- Committee members also mentioned that seed varieties were not always appropriate and felt better technical advice was needed.

Uganda case study

The Uganda case study, from which a number of lessons can be learned, depicts a situation with deficiencies in the joint programming of food and seed resources. Emergency coordination mechanisms, needs assessment and targeting were identified as important issues in joint programming.

Support to IDPs and Refugees in Uganda

Context

Food insecurity has become an increasingly serious problem in Uganda, fuelled by drought, flooding and rebel insurgency. Northern (Gulu and Kitgum) and western (Bundibugyo and Kassesse) districts have been particularly affected by rebel attacks (WFP, 1998b). The majority of the IDPs are living in camps or protected villages, and receive food aid and seeds. Many of these IDPs were surplus-producing farmers before insecurity prevented them from farming. Uganda's refugee population is beginning to be phased out of food assistance. Meanwhile, economic activity and market development flourishes in southern Uganda with entrepreneurs constantly looking for potential market expansion in more secure areas of North and West Uganda. Wholesalers of agricultural inputs and seeds can be found in most secure trading centres. However, only input markets in southern Uganda are well established. Uganda has a reasonably well-developed agricultural sector producing a small surplus of food and seed for local purchase by aid agencies for use in the country and other Great Lakes countries.

Coordination Mechanisms and the Role of Different Stakeholders

Emergency responses are coordinated monthly at the national level. UNHCR hosts meetings addressing refugee issues. An OCHA office has been opened in Kampala to facilitate United Nations inter-agency coordination for relief interventions. District-level responses are coordinated by the District Disaster Management Committee (DDMC). The effectiveness of the DDMC depends upon the capacity of the chairperson (generally a government representative) and on the degree of agency participation in meetings. In cases where district-level coordination is problematic, district-specific meetings are also held in Kampala. WFP is the principal provider of food assistance, operating through implementing partners. The main supplier of seed is the government-based Uganda Seed Project, which produces certified seed of improved and hybrid varieties developed by the NARS at Kawanda and Namulonge.

Seeds and tools are generally distributed by ICRC, NGOs or church organizations, although FAO and WFP have both distributed seeds in response to government requests. WFP has distributed seeds to populations living in the less secure, isolated areas where NGOs are unable to operate.

Needs Assessment and Targeting

Requests for assistance come from the Government or the DDMCs. A United Nations Disaster Management Team (UNDMT) is responsible for needs assessment in the case of new emergencies. UNDMT decisions are based on a joint rapid assessment mission which gathers information on population movement, crop production levels, household food and seed reserves, coping mechanisms, the availability/price of food on local markets and health status.

Additional data generated by WFP's implementing partners and food monitors are used to determine the food requirements of vulnerable populations. Seed requirements are determined by the District Agriculture Officer and collaborating NGOs, on the basis of the success of the cropping season and the availability of seed stocks.

Joint Programming of Food and Seed Resources

Seed and tool distributions are being widely implemented to reduce the need for affected populations to rely on food assistance. Refugees resettled in early 1998 were provided with seeds and tools in March/April as part of an exit strategy in which food assistance will be gradually reduced to zero over a period of four successive cropping seasons. However, post-harvest food assessment missions will determine whether planned reductions in food rations can be implemented. In most cases, seeds are distributed after a WFP food distribution, to avoid consumption of the seed and to encourage planting. In some instances, seeds have been provided to populations not normally targeted by WFP. At the request of Action Contre la Faim, WFP provided seven-day food rations to non-displaced, drought-affected populations of eastern Kitgum District targeted to receive seeds, to avoid seed consumption. Seed consumption was reported when Christian Children's Fund distributed seeds without seed-protection food rations to a drought-affected community. However, World Vision (WV) distributed seed to families hosting IDPs even though these families were not receiving food assistance, as WV considered them to be sufficiently food secure. In Gulu, WFP supports a FFW seed multiplication project implemented by Catholic Relief Services (CRS). Technical assistance is provided by a district agricultural extension officer and a CRS agricultural technician.

Problems Encountered

♦ FAO's guidelines concerning appropriate crops/varieties, seed/tool quality specifications and recommended suppliers have not yet been made available. FAO's technical support to seed programmes is undertaken by an international consultant based in Kenya. Consequently, implementing agencies have no in-country technical advice or coordination for seed and tool programmes. As a result, poor quality hoes and inappropriate sorghum variety mixtures were distributed by NGOs in Kitgum District; there was no technical backstopping for FAO-purchased seed delivered to Gulu and Kitgum districts, resulting in distributions being poorly timed; and poor quality seed has delayed the planned phase-out of food assistance to refugee populations in Adjumani District.

♦ The geographical area and population type (IDPs, host community, drought-affected) to be targeted by NGOs are determined during proposal preparation, without the involvement of the DDMC. This can lead to gaps in the provision of support to some needy populations.

♦ The procurement and distribution of seeds by aid agencies has not adequately considered market impacts, such as the draining of local seed supplies or reductions in the demand for seed from nascent district wholesalers.

♦ Pipeline and security problems have delayed food deliveries to IDP camps by more than one month.

♦ A lack of capacity of local government departments has led to the poorly timed distribution of FAO-purchased seeds in Gulu District and to the distribution of seed to non-targeted, peri-urban populations in Fort Portal, rather than to the targeted, food-insecure IDPs in Bundibugyo District.

♦ In Gulu, inadequate District funding for agricultural extension has limited the amount of technical support provided to the FFW seed multiplication project reducing the quality of seed produced.

4.3 Summary of Issues

When to provide food and seed resources

The complementary provision of both food and seed resources is appropriate when populations are food insecure, and when they do not have their own seed stocks. The duration of food assistance provided and the timing of food provision relative to the distribution of seed depend on: the degree of food insecurity of the target populations; the availability of food and seed supplies; and the degree of security, which will affect the recipients' access to land.

When farming families have lost both their food and seed stocks, it is necessary to provide food assistance immediately and seeds prior to the following cropping season. Food assistance would generally be continued until the newly planted crops are harvested. At harvest, crop yields should be evaluated and estimates made of the number of months the newly harvested food supply will last. This information can then be used to determine whether crop production is sufficient to warrant the phase-out of food assistance or if food assistance should be continued.

Joint food and seed distributions are necessary when:

◊ farmers have suffered total crop loss as a result of conflict or natural disaster;

◊ farmers were displaced as a result of conflict and were not able to harvest their crops;

◊ farmers were unable to sow their crops the previous season owing to emergency-related disruptions;

◊ farmers' food and seed stocks were stolen as a result of rebel attacks; or

◊ IDPs are returning to their homes or refugees are settling on land allocated to them.

In targeting food and seeds, agencies generally prefer to give seeds with food protection rations even when distributing to communities not targeted for food assistance. Under normal conditions, farmers do not eat their seeds. In the case of an emergency, however, implementing agencies have found that when seeds are distributed without food rations, a proportion of seed is often eaten. In Uganda, fears of increased food insecurity led some beneficiaries to wash and consume the seed obtained from distributions. Consumption of small quantities of bean seed distributed by aid agencies is common in Burundi, especially where access to secure land is tenuous.

Not all emergency situations call for the complementary provision of both food and seed. The provision of seeds is not appropriate for food-insecure populations that do not have access to land and therefore have no hope, in the short term, of restoring their own food security through crop production. This situation commonly applies to refugee populations living in transit camps where population density is high. IDPs living in camps with inadequate access to land are also likely to remain reliant on food assistance unless non-agricultural income-generating opportunities are introduced.

As families become more food secure, seeds may be provided without food assistance. Populations that have lost a particular type of crop (for example, following floods or localized pest attacks) may benefit from the provision of seeds to enable them to replant the affected crop and re-establish their own seed supply. In Mozambique, widespread flooding caused the loss of only one of a range of crops that were in the field at the time. UNHCR provided funds for the provision of seed of Matuba, an early maturing maize variety, to flood-affected families, enabling them to plant a second crop of maize during the following dry season and regain their maize seed stocks (World Vision, 1996). Although the affected populations were former refugees, they were considered to be largely food secure. Therefore, food assistance was unnecessary.

The food ration

The relative timing of food and seed distributions was identified as a crucial issue in all countries studied. Food rations before planting help improve the physical fitness of affected families and give them the strength required to prepare their land for planting. This type of situation is common within a one or two-year period after an emergency, when families have begun to re-establish their agricultural activities, but have not yet opened up sufficient land to guarantee food self-sufficiency. Populations that are receiving seeds, but are not receiving regular food assistance, may benefit from the distribution of a seed-protection food ration just before seed distribution. The objective is to encourage beneficiaries to plant the seeds rather than consume them.

Populations that have suffered partial crop loss the previous season and therefore have limited food and seed stocks need less support, but it should be properly focused. Such populations will be short of food during the pre-harvest "hungry period" or "lean period". They should be provided with an appropriate number of months of food rations to bridge the "hungry period" and also with seed prior to planting. The decision of how many months of food rations to provide before planting must be made on the basis of a careful needs assessment, taking into consideration food and income sources and coping mechanisms. In Uganda, food pipeline problems and district-level security problems resulted in delays in the delivery of food aid to IDP camps. Under such conditions, it is crucial that the seeds are **not** distributed at a time when the beneficiaries are waiting to receive food rations, as seed consumption must be expected.

In Burundi and Uganda, WFP sometimes used a 15-day food ration for "seed protection". Such rations are minimal only, providing extra food for a very limited period. These rations are provided mainly to prevent seed consumption and to enable beneficiaries to plant the seeds. If planting is delayed for any reason, for example owing to delays in the onset of the rains, food rations of 15 days may be consumed before planting. In addition, where families fear continuing food insecurity, a 15-day ration may not reduce the pressure to consume seeds. Seed-protection rations of 15 days cannot be expected to have a significant effect on the health status of the beneficiaries or on their ability to prepare their land or maintain their crops during the early stages of crop development. It therefore seems more appropriate to distribute at least a one-month seed-protection food ration to avoid such situations.

When food and seed distributions are implemented simultaneously, it is important that the seed is clearly distinguishable from the food. Preferably, the seed should be dyed, packaged and clearly marked. In some cases, NGOs purchase "seed" from local markets, bring it to the distribution site in 50 kg sacks and distribute it by ladling the "seed" into containers or sacks brought by the beneficiaries. If this type of distribution occurs, it is likely that the beneficiaries will confuse the "seed" received from the NGO with other stored food stocks and consume the "seed" by mistake. This should be avoided whenever possible.

It should be noted that beneficiaries who are given seed are less likely to consume it if they have advance notice of the planned seed distribution and have been encouraged to prepare land for planting. They should also be told that the seeds are being provided so that they can produce their own food, and that food rations will be reduced following harvest. If possible, these explanations should be provided by a local government official rather than by implementing agency staff. Where possible, an agricultural technician should also provide advice concerning optimum crop management practices and the importance of selecting and saving seed.

Planning issues

Effective joint programming of food and seed depends on coordination and capacity. Coordination needs to begin early and was most effective when agencies decided together how and where to intervene. The coordination process should begin before programmes and budgets are fixed. This was the case for the OLS in the Sudan and for Burundi. In Uganda, needs assessment and detailed coordination took place at the district level through the DDMC. However, discussions concerning NGO relief programmes and resource allocations were made at the national level, where inter-agency coordination was weak. As representatives of NGOs participating in the DDMC did not have the authority to commit resources or to change programmes to meet emerging needs, NGO programmes tended to be relatively inflexible. In some cases, this resulted in the incomplete coverage of vulnerable populations.

Coordination is also affected by implementing capacity. For example, in Uganda, insufficient local government capacity led to poorly timed delivery of seeds and, in one case, delivery of seeds to non-targeted communities. Also, a lack of FAO technical advice on demand meant that technical support was limited to periodic visits by an international consultant and NGOs were required to seek their own technical advice. In the Sudan, security problems hindered WFP's logistical capacity and some situations resulted in WFP being unable to provide food rations and seeds according to plan. Limitations in capacity may jeopardize the success of joint distributions and should be considered when implementing partners are selected.

Alternatively, in areas where an agency has a comparative advantage for logistics or implementation this advantage should be used. In Uganda, WFP was able to undertake relief seed distributions and ensure the provision of seeds to communities not reached by other implementing agencies. Similarly in Burundi FAO was able to coordinate information on beneficiary needs, markets and seed quality control for all agencies wishing to implement seed distributions.

Needs assessments and targeting

Innovative approaches to needs assessments and targeting are possible despite the difficulties often inherent in working in a crisis situation. In both Burundi and the Sudan, the household food economy approach helps WFP and its partners assess where joint provision of food and seeds is appropriate. In addition, other information sources such as VAM, FEWS and FAO/WFP crop and food supply assessment missions help identify areas of vulnerability and target resources.

In contrast, in Uganda, despite the availability of various data sources, season-by-season food and seed needs are based largely on district-level estimates collected by implementing agency field staff. For example, although World Vision (Gulu) and Action Contre la Faim (Kitgum) measure crop yields, this information is not always shared with other organizations. Similarly, Investment in Developing Export Agriculture (IDEA) Project publishes information on seed availability that is not regularly used by other agencies when determining seed needs.

Introducing development approaches

In the countries studied, it is hard to ignore a need for rehabilitation and recovery activities given zones of stability exist. These areas have remained outside the conflict for years, but could be brought in at any time. Assistance is needed to meet: basic needs and rebuild livelihoods; the high cost of assistance delivery and the longevity of the crisis, both of which have produced long-term financing concerns, particularly in the Sudan; and, the need to strengthen social and economic capacity, with the aim of laying the foundation to rebuild livelihoods in the event of renewed conflict. In many of these areas aid agencies are unable to maintain a permanent presence because of the threat of insecurity.

Opportunities for introducing development approaches in relief and recovery programmes depend on: the political environment and the degree of security in which programmes are implemented; the willingness of donors to fund development activities; and the innovative capacity of implementing agencies. In Burundi, where reversals in security are frequent, small development aspects have been introduced in joint programmes. For example, one of WFP's implementing partners has created demonstration vegetable plots and provides vegetable seeds at health centres. Similarly, support to a women's group FFW scheme combined vegetable gardening techniques with information on the importance of vegetables for nutrition.

In Uganda, WFP and USAID have begun cooperating in monitoring food and seed market prices to provide information on food market developments and where food or seed aid may hinder developing markets. Although agencies have the opportunity to use this information to help nurture district markets by procuring seeds from district wholesalers instead of from national suppliers, they have not yet done so.

Seed multiplication programmes can be established to increase local seed supplies, multiply selected varieties (local and improved) and facilitate germplasm conservation. These activities require a high level of technical management and a secure environment to be effective. Therefore they can only be undertaken in the later stages of recovery. While seed multiplication can be supported initially through FFW projects (at the pilot stage), this activity should move as quickly as possible to a financially viable enterprise, supported preferably by credit.

FFW may be used to implement a range of agricultural development interventions. It can be used to create pilot farms using improved seed varieties. The food provides participants with the energy to work on the land and the farm promotes agricultural technology dissemination. FFW programmes can also be used to help construct local seed stores, seed banks, grain banks or warehouses, with the objective of maintaining seed supplies and improving food security. Fruit tree nurseries or agro-forestry tree nurseries can also be initiated and maintained through FFW.

4.4 Technical considerations

The Uganda case study highlighted a situation where poor seed quality and inappropriate varieties resulted in the failure of relief seed distributions and delays in the programmed reduction of food assistance to resettled refugees. On the other hand, the Burundi case study showed how the provision of technical guidelines and seed quality control helped guarantee the success of seed relief interventions.

The choice of food and seed distributed depends on a variety of factors:

- ◊ In relief situations, the choice of seed crop often complements the food aid basket. In Burundi, WFP's food basket consists mainly of maize, which is less susceptible to theft and raiding. Simultaneously, FAO has focused seed relief on the principal staple crop - beans - and on vegetables, for their nutritional value and potential to generate income. Although farming families in Burundi grow a wide range of crops, beans are universally grown and represent the preferred food source. Beans also have a short growing season, making it possible to plant three crops a year. They can be grown on limited land areas, therefore they are the most appropriate choice for re-establishing agricultural production. In southern Sudan, staple crops such as maize and groundnut are distributed to help people cope with lean periods and gaps in food assistance.

- ◊ In many cases, implementing agencies may choose to distribute seeds for a range of crops as a way of helping to restore diversity to the farmers' cropping system. In Burundi, NGOs would like to begin distributing a wider range of seeds and/or location-specific crops, such as rice for the low-lying areas. FAO is currently reviewing the options in terms of crops to be distributed. In Uganda, an improved maize variety developed by the Uganda Seed Project is being widely distributed as it produces higher yields while requiring no additional inputs and it fits in with the Government's "Modernization of Agriculture" strategy. Although hybrid maize is commercially available, it is not considered appropriate for seed relief interventions, as seed must be purchased each season.

The provision of guidelines by FAO's Emergency Unit concerning the choice of crop and variety, appropriate seed quality specifications, recommended suppliers, and the regular implementation of seed germination tests prior to seed delivery has helped to ensure that seed distributed in Burundi is appropriate and of good quality. Although similar technical expertise is available in Uganda from the Kawanda Agricultural Research Institute, a lack of agency and donor consensus on the need for technical guidelines has meant that technical inputs are insufficient to ensure both the quality and timing of seed deliveries.

Funding issues

Differences in donor requirements and funding mechanisms can cause difficulties in funding food and seed programmes. For example, the funding of food and seed is often undertaken through different funding channels and ministries. Agencies are therefore required to submit separate funding proposals and meet different requirements.

In Uganda, the European Commission provides food assistance through the European Community Humanitarian Office and funding for seed programmes is limited to relief seed needs, and does not cover developmental activities such as support to developing seed systems. Meanwhile, the United

States Government supports relief activities through the Office of Foreign Disaster Assistance (OFDA) and Title II Food for Peace funds and development activities through USAID, especially with IDEA Project funds. OFDA has specific technical requirements for NGOs undertaking seed programmes, including ensuring food protection rations in food-insecure areas. However, OFDA funds are limited to a year. Beyond this, Title II or other funds must be accessed.

In a typical situation, WFP seeks funding for food for its main caseload, adding an estimation of the food required solely for seed-protection rations. NGOs raise funds separately for seed provisioning. As NGOs are often budget constrained, few are able to undertake programmes if they do not have money in their accounts. Yet donors' regular funding mechanisms have to go through several layers of audit control, which means that once a project is approved, funds may not reach an NGO for a further six months. For this reason, NGOs tend to submit annual project proposals, which limits their ability to respond to unexpected increases in need during the year.

In Burundi, owing to the fluid nature of rebel incursions and insecurity, donors and agencies are more willing to support the pooling of funding for seeds and the provision of food assistance on an ad hoc basis. Although TCOR's funds in support for the Emergency Coordinator last for a year, bilateral donations have already been received to allow FAO to continue to play a coordinating role.

4.5 Appropriate Roles of Major Stakeholders

Where possible, the role of the different United Nations agencies in an emergency should reflect their mandates. WFP would normally assume responsibility for the provision of food assistance. WFP also has a comparative advantage when it comes to logistics, and is able to provide such support (warehouse facilities and/or transport) to other United Nations agencies or to NGOs planning to implement either food or seed distributions.

According to FAO's Mission Statement, "FAO may field an Emergency Coordinator, for up to one year, to coordinate monitoring and assessment activities, provision of agricultural inputs, and planning for recovery and rehabilitation of rural production systems if the nature of the emergency so requires" (FAO, 1997a). In countries where FAO does not have an Emergency Coordinator, other United Nations agencies have become involved in the purchase of seed and its distribution to support populations not targeted by other organizations. For example, WFP was involved in the distribution of seeds purchased by the Ugandan Government in 1998. UNICEF has also distributed seeds and tools in a number of emergencies, including those that occurred in Rwanda, Angola, Burundi, Sierra Leone and the Sudan. It is not clear whether, in such situations, it is more appropriate for other United Nations agencies to become involved in seed relief, or for FAO to ensure that an Emergency Coordinator is in place in countries that require emergency support. Even in situations where FAO does not have an Emergency Coordinator, it should seek to provide in-country capacity for technical advice to organizations implementing agricultural relief interventions.

Appropriate government ministries are usually involved in requesting emergency assistance and in the coordination of emergency interventions. In some cases, they are also involved in implementation. However, a lack of resources and capacity are the main limitations to their direct involvement in food and seed distribution. Successful counterpart relationships between implementing agencies and government staff are relatively common at the national level, where national capacity is the greatest. At the local level, however, local government capacity can be limited and implementing agencies are usually under pressure to deliver relief resources as quickly and efficiently as possible while meeting appropriate standards of accountability. Often the situation-specific circumstances can mean that developing counterpart relationships and building capacity receives much less attention than it deserves. Implementing agencies should work as closely as possible with national and local government counterparts to strengthen their capacity to respond to emergencies in the future.

Some NGOs programme food and/or seed through local partners. Such arrangements offer valuable experience in partnership issues, help to develop indigenous capacity for emergency response and development planning, and encourage community participation in the programming of aid. However, local partners are generally not able to handle large volumes of relief supplies, have limited administrative and reporting capacity, and require close monitoring and control. Nevertheless, United Nations agencies and international NGOs should be encouraged to develop partnerships with local organizations. Working with local NGOs is particularly important as a relief programme becomes reoriented towards recovery, as such partnerships can be instrumental in encouraging beneficiary participation in problem identification and in the definition and development of appropriate responses.

Although some NGOs can have a strong geographical focus, they should be encouraged to "buy-into" national emergency programmes and to provide resources that can be used by themselves in their own area of intervention, as well as by other agencies operating in other provinces. For example, in Burundi, FAO, Christian Aid and the Ministry of Agriculture purchased seed that was distributed by NGOs throughout the country.

The involvement of affected communities in the planning and implementation of relief interventions is generally limited in large-scale emergencies, particularly where the relief programme is well-coordinated at the national level. In Burundi, for example, decisions concerning the crops to be included in the "seed relief package" were made by the FAO Emergency Coordination Unit. Although NGOs expressed interest in diversifying the seed relief package, they sought approval from FAO first before making such modifications. In the case of Uganda, however, where clear guidelines on crop choice were not available, implementing agencies first consulted the beneficiaries concerning their preferences and then confirmed this selection with the District Agriculture Officer and finally with the NARS. Although guidelines are important, they should not be too rigid.

In some cases, beneficiaries have been involved in identifying their basic needs, in developing their own solutions and in carrying out the relief response. Such community participation can result in a locally relevant response with effective, transparent targeting, and also contributes to longer-term development objectives, such as capacity-building, women's empowerment and community empowerment. Community participation in relief interventions is most effective when relationships between the implementing agency and the target communities are well established, and when the agency has a large team of local field staff. The OLS operation in the Sudan is a good example of a large operation working effectively with local communities in situations of inadequate local implementing capacity.

As an emergency situation begins to stabilize, beneficiaries are able to increase their participation in recovery and rehabilitation programmes: in terms of organization, management and commitment; and, to contribute to the activity, for example in the form of labour, local materials, seeds, etc. This increased participation promotes a sense of "community ownership" of the programme. Indeed, beneficiary participation is seen as an essential ingredient to successful implementation and the effectiveness of rehabilitation activities.

Agencies implementing emergency relief and rehabilitation interventions have a responsibility to help build local capacity at all levels: government and national authority counterparts, NGOs, beneficiary communities and the beneficiaries themselves. Unfortunately, the short-term nature of funding for relief and rehabilitation activities makes it difficult to include capacity-building, a long-term activity, as a project objective. Nevertheless, capacity-building should be an unwritten objective of all implementing agencies' activities. New working relationships must be adopted to give more responsibility to the affected populations and their leaders.

5. CONSTRAINTS TO JOINT PROGRAMMING OF FOOD AND SEED RESOURCES

The slow response to support recovery and rehabilitation is in part influenced by decreasing donor budgets and by concerns about the ability to sustain the high level of emergency assistance required to meet basic survival needs. Constraints to providing support for rehabilitation include:

- ◊ instability and the risk of activity reversal. There is also a fear that the creation of assets will encourage attacks by any of the parties engaged in conflict;
- ◊ the lack of continual and timely access to needy areas;
- ◊ the lack of markets and limited trade possibilities;
- ◊ short-term funding arrangements (as mentioned above);
- ◊ weak implementation capacity of counterparts and their intimidation of NGOs and communities; and
- ◊ a weak information base including little socio-economic data with which to identify transition needs and potential responses.

The principal constraints to joint food and seed programming, highlighted by the case studies, are related to: planning, needs assessment and targeting, funding, and technical issues. Major constraints include:

5.1 Planning issues

- ◊ poor coordination of the two resources, particularly when each resource is being provided by a different organization, resulting in their untimely delivery;

- ◊ deterioration in security, logistical problems such as cuts in WFP's pipeline or flooded roads, and lack of transport result in irregular food provisioning and reductions in the size of seed-protection rations, or their cancellation; and

- ◊ limited capacity to implement community-based development activities.

5.2 Needs assessment and targeting

- ◊ poorly coordinated assessment and targeting procedures with different agencies using different approaches; and

- ◊ improperly assessing the needs of communities and providing free food and seeds at inappropriate times, impeding coping strategy recovery and market development.

5.3 Technical considerations

- ◊ inadequate warehouse facilities may result in the need to store food and chemically treated seed in the same warehouse, and inadequate seed and food storage conditions may adversely affect seed and food quality; and

- ◊ limited availability of seed of the most appropriate varieties, which may result in: the distribution of poorer quality seed or of seed of less appropriate varieties; a reduction in the quantity of seed distributed per family; or, a reduction in the number of beneficiaries.

5.4 Funding issues

◊ relief funds are time-limited and funds for longer term activities, such as seed multiplication and testing of new varieties, are more difficult to obtain; and

◊ complications regarding who obtains funds and how quickly reduces responsiveness to needs as there is a very narrow window of time in which to propose projects for funding, receive the funds, and then deliver the resources.

6. OPPORTUNITIES FOR MORE EFFECTIVE AND COMPLEMENTARY PROGRAMMING

The review of the literature and the case studies identified a number of opportunities for improving the joint programming of food and seed resources. These opportunities span a number of technical and financial areas, including programme planning, needs assessments, targeting and funding arrangements and involve all stakeholders. WFP and FAO can strengthen their coordinating relationship, particularly at the field level. Improved coordination will ensure that food and seed assistance are complementary and that beneficiaries optimize the potential of the two resources.

6.1 Planning Issues

◊ Where FAO has an Emergency Coordination Unit, it should take the lead among the UN agencies in coordinating seed relief in response to emergency situations. Where FAO does not have an Emergency Coordination Unit, another means of coordination should be sought. The provision of food rations should be coordinated with WFP.

◊ As WFP often has a comparative advantage in terms of logistical capacity, it is advisable that FAO seeks this support from WFP for the distribution of seeds, particularly in insecure areas.

◊ Emergency distribution plans for the coming season should be discussed with the local coordination committee as soon as possible to ensure effective coordination of the distribution of all complementary relief supplies (food, seeds, tools and non-food items) and to maximize the efficiency of their distribution. Although not always logistically feasible, an attempt should be made to distribute the hand tools before the seeds, to enable the farmers to use the tools to prepare their land for planting.

◊ Where feasible, project proposals should cover more than one cropping season and be as flexible as possible in terms of their geographic focus. In this way:

- funding can be secured for more than one season;
- the joint provision of food and seed resources can be programmed in advance; and
- the second season seeds can be purchased in a timely manner.

◊ Implementing partners should be selected on the basis of prior experience in the delivery of seed and/or food relief. When it is necessary to work with a new implementing partner, including the local Government, it is advisable to consult other donors/agencies working in the area concerning their capacity to implement relief distributions, and to provide close initial monitoring.

6.2 Needs Assessment and Targeting

◊ Where the joint programming of food and seed resources is considered appropriate, an assessment should be made to determine when food resources should be provided to maximize the impact of the seed distribution on the food self-sufficiency of the target populations.

◊ Needs assessment information should be developed at the provincial/district level. United Nations assessment missions should be involved in needs assessment and the definition of appropriate responses. Such missions should assess the needs of the affected populations in terms of food and non-food resources (including seeds and tools) to ensure more effective joint programming of all resources.

◊ There is a need to look objectively at the data being generated by different mechanisms such as WFP's VAM, USAID's FEWS, and NGOs to identify ways of making better use of this information. For example, implementing agencies should be encouraged to measure crop yields at harvest, particularly in food-insecure regions. The data obtained can then be used to give a dynamic picture of changes in regional food supply. By sharing information, agencies can better determine if and when assistance will be required, and for how long.

◊ WFP should continue to be flexible in providing seed-protection food rations to populations not normally targeted to receive WFP food assistance. However, adequate coordination between the seed distributing agency and WFP is essential.

◊ Village relief committees are an important mechanism for improving the effectiveness of assistance. Efforts should be made to avoid duplication of committees and to coordinate their participation in planning, targeting and distribution of all resources.

6.3 Integrating Development Approaches

◊ Implementing agencies should attempt to replace free food and seed distribution as soon as possible with more sustainable activities that strengthen livelihoods and build self-reliance and resilience against future shocks. Examples include a variety of production-based activities, FFW asset creation and rehabilitation activities that satisfy the need for seeds, such as seed exchanges, seed loans, seed multiplication, seed banking, seed sales, and the establishment of seed agents. Seed produced by seed multiplication projects could be purchased by NGOs to satisfy relief needs.

◊ Where NGOs have the capacity to implement agricultural development activities, they should be encouraged to initiate simple agricultural development programmes in parallel to their relief interventions. Appropriate development activities may include: the evaluation of improved agricultural technologies together with farming communities; the initiation of seed multiplication activities; the promotion of fruit and horticultural crops; and the promotion of appropriate seed storage technology. FFW could be used for seed multiplication schemes involving farmers' organizations or women's groups, for the planting and maintenance of fruit-tree nurseries and for the construction of seed stores, grain banks or grain warehouses.

◊ One of the problems that farmers face while attempting to re-establish their agricultural production following an emergency is a lack of opportunity to market their produce. Many NGOs implementing agricultural development programmes work with farmers' organizations to help them sell their produce. WFP could play a role in such programmes by establishing links with such farmers' organizations as part of its local purchase programme.

◊ Activities designed to raise awareness among families attending therapeutic or supplementary feeding programmes of the importance of vegetables in the diet represent a good opportunity for the joint programming of food and seed resources. While therapeutic/supplementary food rations are provided to restore the nutritional status of individuals attending feeding centres, vegetable seeds can be purchased to plant vegetable production demonstration plots and for distribution to care providers as they are discharged from such centres.

6.4 Technical Guidelines and Quality Control

◊ A set of guidelines should be prepared and made available to implementing agencies, specifying appropriate crops/varieties by agro-ecological zone, seed quality specifications for each crop and tool quality specifications. Seed quality specifications should include information on the minimum percentage germination, minimum purity percentage, and maximum moisture content, by crop. These guidelines should be prepared by local experts with input from either the FAO Emergency Coordination Unit or by the designated lead agency, as appropriate.

◊ A list of recommended suppliers of seeds and tools should also be made available to implementing agencies. Suppliers should be recommended on the basis of their performance in terms of timely delivery and the provision of high quality inputs. The list of suppliers should be regularly reviewed and updated, if necessary. Purchasers should be prepared to reject the lowest bid, in cases where quality may be prejudiced.

◊ Phytosanitary certificates provided by seed suppliers as part of their bid for the seed tender do not guarantee that the seeds supplied will meet the seed quality specifications. To ensure that the seed delivered is of good quality:

- procedures to test seed quality (purity, percentage germination and humidity) should be established and appropriate facilities identified;
- seed samples should be taken by the purchasing agency, either during the transfer of the seed from the supplier's warehouse onto their trucks or, if this is not possible, immediately upon delivery in the implementing agency's warehouse; and
- seed samples should immediately be sent to the recommended seed testing facility(ies) for purity, germination and humidity tests.

◊ All agencies purchasing seed for relief interventions should be encouraged to include the following in their seed contracts/tenders:

- a **delivery date** for each destination;
- a clause that states that they have the right to carry out an **independent evaluation of seed quality** either during the transfer of the seeds from the supplier's warehouse to the trucks or upon delivery; and
- penalties for late delivery and poor seed quality.

◊ When certified seed is available, implementing agencies should be encouraged to purchase it rather than commercial seed. When certified seed is not available, commercial seed rather than uncleaned food grain should be purchased.

◊ Implementing agencies should be encouraged to place their seed orders with the supplier as early as possible, to guarantee delivery of the correct quantity of good quality seed and thus enable joint food and seed operations to be carried out successfully. Relief agencies should consider contracting seed companies or professional growers to multiply specified quantities of seed of selected varieties proven to be adapted to the target environment.

◊ Agricultural research organizations should attempt to generate more information concerning low-input agricultural systems as a disaster preparedness and mitigation strategy. Whenever possible, this work should be done in collaboration with NGOs implementing agricultural development programmes, as a way of exchanging information and sharing research results.

6.5 Funding Mechanisms

◊ Recovery funds should be made available to support livelihood systems and the re-establishment of local and national seed supply systems. Such activities could be carried out as part of a relief/recovery proposal, or as part of a parallel development programme.

◊ Special relief funds, which can be accessed relatively quickly, are particularly useful to the joint programmeming of food and seeds. While these funds are limited to relief interventions they enable agencies to supply both food and seeds in a timely manner.

◊ More flexible and longer-term funding mechanisms will allow implementing agencies to incorporate both emergency relief and simple "development" activities, such as seed multiplication schemes in their funding proposals.

6.6 Monitoring and Evaluation

◊ The complementary distribution of food and seed resources can be appropriate during all phases from relief to development. Simple monitoring of joint food and seed operations should be carried out to determine whether the resources have been effectively used and to identify ways to improve ongoing and future interventions.

◊ Monitoring and evaluation must go beyond simply monitoring the logistical performance (quantity of food distributed, number of beneficiaries receiving food and/or seeds), and examine the effect of food and seed resources on the target communities, considering issues of timing, appropriateness of the food ration, seed varieties, and distribution mechanisms.

7. REFERENCES

ACORD (1995). *Development in conflict: The experience of ACORD in Uganda, Sudan, Mali and Angola*. Relief and Rehabilitation Network Paper No. 9. Overseas Development Institute, London.

Borton, J., Nicholds, N., Shoham, J., and Mukhier, O.J. (1992). *An evaluation of CONCERN's 1990-1991 emergency programme in Kosti Province, Sudan*. Overseas Development Institute, London.

Boudreau, T. (1998). *The food economy approach: a framework for understanding rural livelihoods*. Relief and Rehabilitation Network, Paper No. 26. Overseas Development Institute, London.

Bush, J. (1995). *The role of food aid in drought and recovery: Oxfam's North Turkana (Kenya) Drought Relief Programme, 1992-94*. Disasters Vol 19, No. 3.

Campbell, W., and Harvey, P. (1998). *Food-assisted programmeming in the Greater Horn of Africa.* A report to CARE. London. (mimeo)

CARE (1996). *CARE food and livelihood security strategy for the Greater Horn of Africa.* Planning grant proposal (draft), CARE, Atlanta, May 1996. (mimeo)

Chapman, J., White, J., and Nankam, S. (1997). *World Vision's experience with seed supply during emergency and resettlement programmes in Mozambique and Angola: Implications for the future.* Paper presented at "Enhancing research impact through improved seed supply: Options for strengthening national and regional seed supply systems", March 10-14, 1997. Workshop sponsored by ICRISAT, ICARDA, IITA and GTZ.

FAO (1997a). *FAO's emergency activities.* Rome.

FAO (1997b). *La Production des Semences de Qualité Déclarée au Rwanda.* Rome.

FAO (1998). *Develop institutional agreements and capacity to assist farmers in disaster situations to restore agricultural systems and seed security activities.* Report of a project planning meeting held 2-4 February, Rome, Italy. Seed and Plant Genetic Resources Service, Plant Production and Protection Division, FAO, Rome.

Friis-Hansen, E., and Rohrbach, D.D. (1993). *SADC/ICRISAT 1992 drought relief emergency production of sorghum and pearl millet seed: Impact assessment.* ICRISAT Southern and Eastern Africa Region Working Paper 93/10.

Grunewald, F. (1996). *Food aid: For or against?* International Review of the Red Cross, No. 312. (October).

Howell, P. (1997). *Crop failure in Dalocha, Ethiopia: ActionAid's participatory emergency response.* Relief and Rehabilitation Network Newsletter No. 8, (May).

Macrae, J., Bradbury, M., Jaspars, S., Johnson,D., and Duffield, M. (1997). *Conflict, the Continuum and Chronic Emergencies: A Critical Analysis of the Scope for Linking Relief, Rehabilitation and Development Planning in Sudan. Journal of Disaster Studies, Policy and Management, 21 (3): 223-243.*

McDonagh, M. (1996). *Une méthodologie d'économie alimentaire pour évaluer les moyens d'existence.* WFP, Burundi.

Mourey, A. (1997) *Le Comité international de la Croix-Rouge face aux problèmes de nutrition des victimes de la guerre.* Food, Nutrition, and Agriculture, 19:37-41.

Mukupo, D. (1994). *Responding to the 1991/92 drought in Zambia: The programmee to prevent malnutrition (PPM).* Relief and Rehabilitation Network, Paper 2. Overseas Development Institute, London.

OCHA, (1998). *United Nations Consolidated Inter-Agency Appeal for Angola.* New York and Geneva.

ODI, (1996). *Relief and Rehabilitation Network Good Practice Review 4. Seed provision during and after emergencies.* ODI Seeds and Biodiversity Programme, December 1996. Overseas Development Institute, London.

Peppiat, D., and Mitchell, J. (1997). *Buying power: the case for a cash-based response to famine.* Discussion paper on the use of cash or vouchers as food entitlement systems. International Division, British Red Cross.

Pottier, J. (1996). *Agricultural Rehabilitation and Food Insecurity in Post-war Rwanda: Assessing Needs, Designing Solutions.* In Swift, J. (ed.), *War and Rural Development in Africa.* IDS Bulletin, 27(3):56-76.

Richards, P., and Ruivenkamp, G. (1997). *Seeds and Survival: crop genetic resources in war and reconstruction in Africa.* IPGRI, Rome.

Sitch, L. (1996). *Fate of seed/planting material of farmer selected varieties distributed through World Vision's extension programme.* World Vision Mozambique, Maputo.

Sitch, L., Bias, C.A., Henderson, S.A., and White, J.G. (1996). *Multi-location testing of improved varieties of maize on farmers' fields in northern and central Mozambique.* CIMMYT Proceedings of the Fifth Eastern and Southern Africa Regional Maize Workshop, Arusha, Tanzania, June 3-7, 1996.

Sperling, D., Sitch, L.J., White, G., and Zandamela, C.B., (1994) *The evaluation of open pollinated maize varieties under a range of zero input farming systems in central Mozambique.* CIMMYT Proceedings of the Fourth Eastern and Southern Africa Regional Maize Conference, *Maize Research for Stress Environments*, Harare, Zimbabwe. 28 March - 1 April 1994. pp. 101-105.

Sperling, L. (1996). *Rwanda nationwide household survey on the status of the bean, sorghum and manioc crops 1995: the impact of war on production and varietal erosion.* Seeds of Hope. The Hague, The Netherlands.

Sperling, L., (1997). *The effects of the Rwandan war on crop production and varietal diversity: A comparison of two crops.* Agricultural Research and Extension Network Paper 75. Overseas Development Institute, London.

USAID (1996). *Seeds for disaster mitigation and recovery in the Greater Horn of Africa.* Report prepared for USAID by the DESFIL project of USAID/G/ENV/ENR by Chemonics International, Inc. Washington, DC.

WFP (1997). *Evaluation des besoins alimentaires en 1998: Burundi.* World Food Programme, Burundi.

WFP (1998a). *From crisis to recovery.* Policy Issues: Agenda item 4. Paper prepared for submission to the Executive Board Annual Session, Rome, 18-21 May. World Food Programme, Rome.

WFP (1998b). *Uganda:* Country *strategy outline.* World Food Programme, Uganda.

WFP (1998c). *Sudan Case Study.* World Food Programme, Rome.

World Vision (1996). *Emergency Maize-Pak Distribution, Tete, Sofala and Zambézia Provinces.* Report of the UNHCR emergency distribution of maize-paks to flood-affected families in 1996. (mimeo)

RESTORATION OF SEED SYSTEMS AND PLANT GENETIC RESOURCES AFTER DISASTERS: A Synthesis of the Background Papers

Victor N. Bushamuka
P.O. Box 76, Bradfield Hall,
Cornell University
Ithaca, NY 14853, USA

1. INTRODUCTION

It is widely recognized that the maintenance of genetic diversity is essential for the sustainability of local farming systems and agricultural production. In this regard, agricultural systems in most developing countries, and especially disaster-prone areas, are characterized by a wide range of crops and varieties necessary to achieve food security and to meet other diverse needs of populations living in rural communities. Therefore, farmers access to seed of diverse local varieties has been, and will continue to be, the foundation of present and future agricultural systems in developing countries.

Most farmers in developing countries still largely depend on on-farm production of seed and plant material. These agricultural systems have been time-tested and are found to be stable. However, they are also extremely fragile during times of disaster. Generally, countries and farming communities are unprepared for disasters and their subsequent ramifications. Farmers' seed stocks can be destroyed, stolen or consumed, and seed of locally adapted varieties may be lost entirely. Furthermore, countries and communities where disasters occur frequently lack the capacity to restore their seed supply systems. Due to the lack of preparation for emergency seed relief and the complexity and difficulty related to emergency operations, most relief programmes have been unable to restore the majority of the adapted local varieties lost due to disasters.

In an effort to restore seed and plant genetic resources after disasters, a project entitled "Developing Institutional Agreements and Capacity to Assist Farmers in Disaster Situations to Restore Agricultural Systems and Seed Security Activities" was funded by the Norwegian Government. The project objectives are as follows:

(i) to develop mechanisms that enable countries and their rural farming communities to identify, acquire, multiply, and deliver seed of locally adapted varieties before, during, and after disasters; and

(ii) to determine mechanisms of collaboration for a continuing partnership among governments and institutions involved in the maintenance or restoration of farmer seed systems, plant genetic resources, and seed supply systems.

In addition, the project aims to identify follow-up actions to achieve these objectives and to determine the role of concerned institutions.

The project called for two meetings, the first of which was organized in January 1998 as a planning session. The purpose of the second workshop of November 1998 was to convene experts from around the world to identify appropriate strategies for achieving the above mentioned project objectives. In preparation for this workshop, six background papers that analyzed the pertinent issues relating to the restoration of adapted local varieties were prepared. These papers include the following: (i) characterizing disasters; (ii) farmer seed systems and disaster; (iii) plant genetic resource and seed relief; (iv) regulatory aspects of seed security; (v) seed stocks and seed multiplication in emergency situations; and (vi) food and seed assistance in the recovery from crisis.

The authors of the papers have stressed the importance of local varieties and crop diversity in agricultural systems of disaster-prone countries. They have also emphasized that in order to design strategies for the restoration of local varieties after disasters, it is first important to understand why farmers perceive the importance of crop genetic diversity and how they maintain it in the farming systems.

2. FARMING SYSTEMS AND CROP GENETIC DIVERSITY

In most developing countries, farmers living in disaster-prone areas attach great importance to yield stability for food security, which depends on accessibility to genetic diversity in the form of crops and varieties. This stability is the result of repeated natural and farmer selection pressure under local conditions, mutation, and uncontrolled hybridization. Growing mixed varieties and crops reduces insect and disease pressure that could drastically affect yield. In addition, different crops and varieties provide an opportunity to make better use of available resources as the difference in planting and harvest dates spreads the labor demands for plowing, planting, weeding and harvesting. This practice is viewed by farmers as a way to reduce the impact of short-term disasters. However, producing and maintaining a wide range of crops and varieties requires the availability of land.

Socio-cultural and socio-economic factors also contribute to the maintenance of diversity in farming systems. These may include gender specific crop and variety preferences. For example, men and women farmers may differ in their selection of crop variety according to the transformation and storage qualities. Cultural norms may also influence the selection of crop varieties produced for consumption. For instance, most landraces are low yielding, but are nonetheless maintained in farming systems throughout the developing countries for their specific cultural or culinary values. In the Great Lakes region, despite their low yields, some sorghum and banana varieties are maintained in farming systems specifically for the production of local beer. However, when production is under stress, limited resources farmers often favor varieties that guarantee food stability at the expense of those grown for culinary or construction qualities.

Production of and accessibility to new and adapted varieties also contributes to the increase of crop genetic diversity in farming systems. Farmers acquire a new variety by procuring seed from formal sector seed producers and maintain the variety in the following cropping seasons if the farmers' expectations are satisfied. There are also farmers who prefer to acquire seed of new varieties from neighbors and relatives rather than from the original source. In this case, the new variety is spread from farmer to farmer.

In summary, it is important to stress that crop genetic diversity, - - the basis of agricultural systems in the developing countries - - is mainly maintained through farmer seed systems.

3. RESTORING SEED SYSTEMS AFTER DISASTERS

The authors of the background papers stress that in order to restore the agricultural systems after disasters, it is critical that farmer seed systems are re-established. However, due to the complexity of the seed systems, the multiple difficulties faced by emergency operations, and the maintenance of seed diversity depends on how well the institutions involved in seed security activities and plant genetic resources are prepared for disaster situations.

This document will highlight approaches elaborated in the background papers to identify, acquire, multiply, and deliver seed of adapted local varieties and describe the institutional collaboration needed for each of the above steps.

3.1. Identification of appropriate varieties

In order to restore varieties lost during disaster situations, it is first imperative to know the crop germplasm used by farmers. It is therefore necessary that varieties used by farmers in disaster-prone areas are characterized and evaluated and that the information is made available to disaster relief operations. Presently, we are far from achieving this goal. There is little information available on most varieties used by limited resource farmers in disaster-prone countries. The information that does exist is collected by different institutions, which makes it difficult for emergency seed relief providers to obtain. Furthermore, it is often poorly documented or incomplete. For example, collected accessions in genebanks frequently lack socio-economic data critical to the restoration of farmer varieties. The authors of the papers have underscored that the availability of information on varieties used by farmers and their management characteristics in disaster-prone areas is the first major step in the preparation for disaster situations.

To facilitate the identification of appropriate varieties for restoration in disaster situations requires work towards the following:

(i) improving the information exchange and documentation on the relevant varieties that have been characterized and evaluated;

(ii) characterizing and evaluating varieties used by farmers and creating lists of names of local varieties; and

(iii) producing variety maps that are combined with socio-economical and cultural information relevant to the management of farmer seed systems.

There is a wide range of organizations that can be consulted and mobilized to provide relevant information on local varieties. These include among others national and international research institutions, private and public institutions, international institutes and United Nations agencies, seed and plant genetic resources regional networks, Non-governmental organizations (NGO), and farmer organizations.

3.2. Acquisition of appropriate varieties

After the appropriate varieties for the disaster area have been identified, the following major task is to transport the varieties back to the affected region. This issue was addressed in detail by the background papers on farmer seed systems, seed regulation aspects of seed security and the seed stocks and seed multiplication in emergency situations. The authors of the documents show that local markets, neighboring countries or regions, and national and international genebanks and collections constitute the major sources of seeds of local varieties in emergency situations. In cases when the appropriate varieties are located in neighboring countries, seed regulations may prohibit the movement of seeds across borders, especially for local varieties.

Countries have seed regulations including seed laws (variety release and seed certification), phytosanitary laws and intellectual property rights systems which often interfere with the acquisition of seed varieties. Regulations, especially in the case of local varieties, which are often not released or registered, can seriously hamper the restoration efforts. For example, in some countries seeds of non-released varieties are not allowed in the markets, which can hurt farmers during emergency situations when larger quantities are often needed. In addition, multiplication of seeds of non-released varieties collected from affected countries can be illegal in neighboring countries.

Although the majority of farmers' seeds are local varieties, large numbers of farmers in disaster-prone countries use certified seeds or commercial seeds of released varieties. In emergency situations, these seeds can be acquired from national and international seed producers or in the neighboring countries or regions. However, once again regulations on seed importation and exportation can hamper the acquisition of appropriate seeds. These importation and exportation regulations affect both certified and non-released local varieties.

It is suggested that regulation issues in emergency situations, which are the responsibility of each government, must be addressed prior to disasters. In many countries, seed regulations are created by members of a national seed board or council who are often involved in the formal seed sector. As a result, the individuals making seed laws are rarely directly involved in discussions on the conservation or restoration of local farmer varieties after disasters. The success of adapting appropriate regulation strategies during disasters depends heavily on the collaboration of governments and the institutions involved in seed security and plant genetic resources. These institutions include non-governmental organizations, private and public seed producers, national and international plant genetic resource institutions, farmer organizations, and regional seed security networks. Collaboration needs to be established at national, regional, and global levels.

3.3 Seed multiplication in crisis

When seeds of appropriate varieties have been acquired, adequate quantities of seed are needed to restore seed systems and local crop genetic diversity in communities where disasters have occurred. The National Agricultural Research Systems (NARS) and/or International Agricultural Research Centers (IARC) are the major providers of seed for multiplication in most disaster-prone countries. However, while these research institutions develop and test varieties, they do not have the mandate or the resources to multiply and store large quantities of seeds of adapted varieties.

In emergency situations, seeds of local varieties can be procured from local markets or in neighboring countries or regions with similar agro-ecological conditions. These seeds are often purchased as grain and, as a result, the quality is not guaranteed. While national and international genebanks and collections have considerable sources of seeds that have been characterized and stored, these seeds are kept in small quantities that require extensive multiplication efforts to be used in emergency.

The background paper on seed stocks and seed multiplication in emergency situations clearly shows that seed stocks are costly and often avoided by governments in developing countries. Therefore, the appropriateness of seed stocks must be carefully analyzed before being suggested as a strategy for the restoration of farmer varieties.

Only national and international seed producers have the capacity to multiply and store adequate quantities of seeds. Furthermore, these producers often keep stocks of unsold certified seeds. However, companies are mostly involved with commercial crops, which most often do not include the varieties needed by limited resource farmers (seeds of commercial crops represent approximately 5 to 10% of farmer seed requirements).

Ways in which this problem might be addressed are needed. In addition, linkages between genebanks, NARS, IACRs, seed companies in the region, NGOs, farmer organizations, and regional seed security networks must be established.

3.4 Seed delivery in disaster situations

Seed delivery has a stepwise process, which includes planning, needs assessments, funding, and relief operations.

Planning for emergency seed intervention starts with the characterization of farmer varieties and their system of production, the identification of potential places for multiplication in times of calamities, and agreements for collaboration from the major stakeholders. During disasters joint planning for food and seed emergency intervention is required to coordinate the appropriate timing for seed and tool distribution, determine the number of cropping seasons for funding, and determine whether food for seed protection is needed.

Disasters are very diversified and take many different forms. The main factors that need to be assessed when determining the effects of disasters on farmer seed include the timing of the disaster and its duration and frequency. Needs assessments include a consideration of the crops required, the phenological and ecological characteristics required in the affected areas, and socio-economical factors that are relevant to the farming systems and on-farm seed supply systems. Furthermore, changes in social and economical conditions due to disasters may lead to new production systems. When this happens, new varieties that vary slightly to the pre-disaster varieties may be appropriate. Technical considerations such as the availability of adequate warehouse facilities are also critical to the success of seed delivery and should be considered when assessing needs. For example, limited warehouse availability may result in the storage of food and chemically treated seed in the same warehouse. Inadequate seed storage conditions can affect seed quality and thus reduce the availability of seed of the most appropriate varieties. This can lead to a decrease in the quantity of seed per farmer or per number of beneficiaries and to the distribution of less adapted varieties.

Funding for relief activities is time-limited and as a result important activities such as seed multiplication and testing of new varieties in crisis situations do not receive adequate financial support. There is a narrow window of time to propose projects, receive funds, and deliver resources, which even under normal circumstances is often a slow and complicated process, thus restricting emergency relief efforts.

Stakeholders in the provision of seed following a crisis are the affected communities, the governments where the disaster occurred, United Nations Agencies, donor agencies, international and national NGOs who usually act as implementing partners to UN organizations, and regional seed security and plant genetic resources networks.

4. INSTITUTIONAL COLLABORATION

The papers prepared for this workshop illustrate that linkages between governments and institutions involved in seed production and seed security activities, emergency food and seed provision, and management of plant genetic resources are critical to the restoration of seed and plant genetic resources.

It is necessary to determine how the various players can collaborate to identify local varieties used by farmers in disaster-prone areas and to establish mechanisms to facilitate the acquisition of appropriate varieties, multiplication and delivery of seed to the affected areas. Frameworks or conditions to promote collaboration between institutions must also be identified.

Each step in the process of restoring local varieties lost to disasters is extremely complex. There is always a high number of institutions that are involved in activities related to seed security and plant genetic resources, and the collaboration of these institutions to the restoration effort is critical in any given region. Coordination of disaster preparedness, relief and rehabilitation is needed at local, national, regional and global levels and the leading organization should act as a liaison between the different institutions involved. However, the success of a new initiative such as this requires adequate and ongoing resources.

5. BIBLIOGRAPHY

FAO. 1998. Seed stocks and seed multiplication in emergency situations. International workshop on developing institutional agreements and capacity to assist farmers in disaster situations to restore agricultural systems and seed security activities. 3-5 Nov. 1998 Rome, Italy.

Grunewald, F. 1998. Characterizing disaster. International workshop on developing institutional agreements and capacity to assist farmers in disaster situations to restore agricultural systems and seed security activities. 3-5 Nov. 1998 Rome, Italy.

Hines, D., and S. Wickrema, and L. van Straaten. 1998. Food and seed assistance in the recovery from crisis. WFP. International workshop on developing institutional agreements and capacity to assist farmers in disaster situations to restore agricultural systems and seed security activities. 3-5 Nov. 1998 Rome, Italy.

Hodgkin, T., and N. M. Anishetty. 1998. Plant genetic resource and seed relief. International workshop on developing institutional agreements and capacity to assist farmers in disaster situations to restore agricultural systems and seed security activities. 3-5 Nov. 1998 Rome, Italy.

Longley, C. 1998. Farmer seed systems and disaster. International workshop on developing institutional agreements and capacity to assist farmers in disaster situations to restore agricultural systems and seed security activities. 3-5 Nov. 1998 Rome, Italy.

Louwaars, N. , and R. Tripp. 1998. Regulatory aspects of seed security. International workshop on developing institutional agreements and capacity to assist farmers in disaster situations to restore agricultural systems and seed security activities. 3-5 Nov. 1998 Rome, Italy.

Musa, T. 1998. Farmer seed systems. International workshop on developing institutional agreements and capacity to assist farmers in disaster situations to restore agricultural systems and seed security activities. 3-5 Nov. 1998, Rome, Italy.

ANNEX I

LIST OF PARTICIPANTS

BISHAW, Zewdie
Seed Systems Specialist
Seed Unit, ICARDA
West Asia and North Africa Seed Network
P.O. Box 5466
Aleppo, SYRIA
Fax 963 21 2213490
E-mail ICARDA@CGIAR.ORG

BRAMEL-COX, Paula
Acting Director, Genetic Resources and Enhancement Programme
ICRISAT
Patancheru 502 324
Andhra Pradesh, INDIA
E-mail ICRISAT@cgnet.com

BURGAUD, Francois
GNIS
44 Rue du Louvre
75001 Paris, FRANCE
Fax 33 1 4233 2774
E-mail burgnis@club-internet.fr.

CHAYA, Mona
Co-ordinator of Agricultural Intervention
OSRO/TCOR
Obala Kulina Bana 4
71000 Sarajevo, BOSNIA HERZEGOVINA
E-mail mchaya@unops.undp.com.ba

CUSTOVIC Hamid
Agricultural Faculty – Sarajevo
sl. Zmaya od Bosne 10
71000 Sarajevo
Tel. 387-71- 653 033 / 231 188
Fax 387 71 667730
E-mail FAO_BH@ZAMIR-SA.ZTN.APC.ORG

DASHIELL, Ken
IITA Acting Director
Crop Improvement Division
c/o L.W. Lambourn and Co.
Carolyn House 26 Dingwall Road
Croydon CR9 3EE, ENGLAND
Fax 234 2 241 2221
E-mail k.dashiell@cgnet.com

DEMISSE, Abebe
General Manager
Biodiversity Conservation & Research Institute
P.O. Box 30726
Addis Ababa, ETHIOPIA
Fax 251-1-61-37-22
E-mail Biod.et@telecom.net.et

DIOP, Amadou
Directeur de l'Agriculture
Ministère de l'Agriculture et de l'elevage
B.P. 323
Niamey, NIGER
Tel. (227) 752335

ELANG Joseph
ISI, Cabinet du Ministre
Ministère de l'Agriculture
Yaounde, CAMEROON

EYZAGUIRRE, Pablo
IPGRI
Via delle Sette Chiese, 142
00145 Rome, ITALY
Fax (39-6) 5750309
E-mail IPGRI@CGNET.COM

FORNASARI. Luca
Clos de l'Ermitage
636 Av. E. Jeanbrau
34090 Montpellier, FRANCE
E-mail LGFORNASARI@WANADOO.FR

FRANCISCO, Castame Antonio
Vice Chairman, National Committee of Plant Genetic Resources
Ministry of Agriculture and Rural Development
Prédio C.N.I.C., C.P. 10825 (BG)
Luanda, ANGOLA
E-mail fitogen@ebonet.net

GALIONE Caterina
Associazione Sementieri Mediterranei
(AS.SE.ME)
Via Parigi 11
00185 Rome, ITALY

GASCON, Jean-François
Co-ordinator
TCOR/FAO
Kigali, RWANDA

GRAHAM, Lyle Barbara
Inter American Institute for Co-operation on Agriculture (IICA)
Representative Waterfront Castries
BOX 1223
St. Lucia, WEST INDIES

GRUNEWALD, François
Groupe Urgence-Rehabilitation-Developpement
La Fontaine des Marins
26170 Plaisans, FRANCE
Tel/Fax 0033 04 75282936
E-mail f.grunewald@infonie.fr

HAWTIN, Geoffrey
Director-General
IPGRI
Via delle Sette Chiese, 142
00145 Rome, ITALY
Fax (39-6) 5750309
E-mail IPGRI@CGNET.COM

HEFFER, Patrick
FIS/ASSINSEL
Chemain du Reposoir 7
CH-1260 Nyon, SWITZERLAND
E-mail Fis@iprolink.oh

HODGKIN, Toby
Principal Scientist
IPGRI
Via delle Sette Chiese, 142
00145 Rome, ITALY
Fax (39-06) 5750309
E-mail IPGRI@CGNET.COM

KANO, Aminu Mutar
Federal Ministry of Agriculture and Natural Resources
Abuja, NIGERIA

LANDI, Renzo
Professor
University of Florence
Via G. Mameli, 19
50131 Florence, ITALY
E-mail rlandi@dape.agr.unifl.it

LEPHEANA, R.M.N.
Director of Crops
Ministry of Agriculture
Department of Crops
Maseru, LESOTHO

SHI, Lin
Second Secretary
Permanent Representation of P.R. China to FAO
Via della Caffarella, 9
00179 Rome, ITALY

LONGLEY, Catherine
Research Fellow
Overseas Development Institute (ODI)
Portland House
Stag Place
London SW1E 5DP, UNITED KINGDOM
Tel 44 171 393 1600
Fax + 44 171 393 1699

LOWAARS, Niels
CPRO-DLO
P.O. Box 16
6700AA Wageningen
THE NETHERLANDS
Fax 31 317 418049
E-mail n.p.louwaars@cpro.dlo.nl

LUCCONI, Enrico
Associazione Sementieri Mediterraneo
(AS.SE.ME)
Direttore
Via Parigi, 11
00185 Rome, ITALY

MACHIRI, Shiela Tatenda
Deputy Sector Co-ordinator
SADC Food Security Unit
P.O. Box 4046
Merchant House, 43 Robson Manyika Avenue
Harare, ZIMBABWE
Fax 263 4 795345
E-mail S.Machiri@SADC.org.zw

MAK, Soeun
Chief of the Technical Office
Department of Agronomy
Ministry of Agriculture, Forestry and Fisheries,
(c/o FAO #5 Street 370)
Phnom Penh, CAMBODIA
Tel 855-15-837226
Fax 855-23-216547

MAMICPIC, Noel Gayo
Director
Asia & Pacific Seed Association (APSA)
P.O. Box 1030
Kasetsart Post Office
Bangkok 10903, THAILAND
Fax 66 2 940 5467
E-mail apsa@apsaseed.com

MEYER, Raymond
U.S. Office for Foreign Disaster Assistance
8.06.044, RRB
USAID
Washington, DC 205223-8602, USA

MONDE, Sama
Director
Rice Research Station (RRS)
Rokupr, PMB 736
Freetown, SIERRA LEONE
Fax (232 22) 229545
Tel (232 22) 223282
Email rokupr@sierratel.sl

MULUMBA, John Wasswa
Chairman EAPGREN
c/o NARO
P.O. Box 295
Entebbe, UGANDA
Fax 256 42 21070
E-mail narohz@imul.com

MUSA, Temba Miziyabo
Seed Provision Consultant, Seed Provision Systems
8 Patula Av. Borrowdale
Harare, ZIMBABWE
Fax 263 4 88 2987
E-mail tmusa@interzim.icon.co.zw

NANKAM, Claude
Director, Agricultural Programmes, Angola
World Vision International
Rua João de Barros
Luanda, ANGOLA
E-mail Claude_Nankam@wvi.org

NARDI, Marco
Associazione Italiana Sementi
Segretario
Piazza della Costituzione 8
40128 Bologna, ITALY

OMALIKO, Chukwuemeka Prosper Eprahim
Director
Federal Ministry of Science and Technology, Agric.Natural Sciences and Dept. Technology
P.M.B. 331 Abuja, NIGERIA
Fax 234-9-5233903/52355204
E-mail gpo.abuja@nipost.pinet.net.

PEACKOCKE, Barnaby
Action Contre la Faim
Head, Food Security Unit
4, Rue de Niepce
75014 Paris, FRANCE
Tel 33 01 4335 88 88
E-mail so@nef.imaginet.fr

PERRET, Pierre Michel
ICRC Relief Division
19, Avenue de la Paix
CH 1202 Geneva, SWITZERLAND
Fax 022 733 2057

QUAGLIOTTI, Luciana
Universitá di Torino
Via P. Giuria, 15
10124 Torino, ITALY

REMINGTON, Tom
Catholic Relief Services – East Africa
P.O. Box 4967
Nairobi, KENYA
E-mail agric-ea@form-net.com

RICHARDS Paul
Wageningen Agricultural University
De Niewolanden
6709 PA Wageningen, THE NETHERLANDS
Tel + 31 (0) 317 485031
Fax + 31 (0) 317 484759
E-mail paul.richards@tao.nl

SEIDU, Mellon
Embassy of the Republic of Ghana
Via Ostriana, 4
00199 Rome
Te.l 86215671

SCOWCROFT, William
Director, Agriculture Australia Consultants
P.O. Box 154
Horshan, Victoria, 3402 AUSTRALIA
Tel +61 3 5362 2345, Fax + 61 3 5362 2348
E-mail SCOWCROFTW@VIDA.AGVIC.GOV.AU

SPERLING, Louise
Senior Scientist
CIAT
Frankenstraat 26, 2582 SL
The Hague, THE NETHERLANDS
E-mail L.Sperling@cgiar.com

TAFESSE, Kebede
Head, Planning and Programmeming Service
National Seed Industry Agency
P.O. Box 9197
Addis Ababa, ETHIOPIA
Tel 251 1 530982
Fax 251 1 519051
E-mail nsia@mail.telcom.net.et

TOLL, Jane
IPGRI
Via delle Sette Chiese, 142
00145 Rome, ITALY
Fax (39-06) 5750309
E-mail IPGRI@CGNET.COM

TUNWAR, N.S.
Senior Technical Advisor
AFG/96/004 FAO Afghanistan
F6/9, 5T45, HN07
Islamabad, PAKISTAN
Tel (92-51) 828217-882818-826434
Fax (92-51) 826439
E-mail sharif@isle.comsats.net.pk

WALKER, David John
Natural Resources Institute
The University of Greenwich
Chatman Maritime, Kent ME44TB, UNITED KINGDOM
Tel 44 1634 883760
Fax 44 1634 883567
E-mail D.J.Walker@gre.ac.uk.

WICKREMA, Sonali
WPF
Rome, ITALY
E-mail sonali.wickrema@wpf.org

FAO

Mailing address: Vialle delle Terme di Caracalla
00100 Rome Italy

ANISHETTY, Murthi
Senior Officer
Seed and Plant Genetic Resources Service
Tel: +39-0657054652
Fax: +39-0657056347
e-mail: murthi.anishetty@fao.org

BUSHAMUKA, Victor
Seed and Plant Genetic Resources Service
Tel: +39-0657053751
Fax:+39-0657056347

FIEBIG, William
Agricultural Officer
Seed and Plant Genetic Resources Service
Tel: +39-0657053517
Fax: +39-0657056347
e-mail: william.fiebig@fao.org

HINES, Deborah
Sr. Natural Resource Economist
FAO – TCIP
Rome, Italy
E-mail Deborah.Hines@FAO.org

LARINDE, Michael
Agricultural Officer
Seed and Plant Genetic Resources Service
Tel: +39-0657053969
Fax: +39-0657056347
e-mail: michael.larinde@fao.org

MENINI, Umberto
Chief
Seed and Plant Genetic Resources Service
Tel: +39-0657054027
Fax: +39-0657056347
e-mail: umberto.menini@fao.org

SIKORA, Ivan
Senior Officer
Seed and Plant Genetic Resources Service
Tel: +39-0657053485
Fax: +39-0657056347
e-mail: ivan.sikora@fao.org

THOMAS, Laurent
Operations Officer
Service for Special Relief Operation (TCOR)
Tel: +39-0657055042
Fax: +39-0657054941
e-mail: LaurentJacques@fao.org

VINET, R.A. Rodrique
Operations Officer
Service for Special Relief Operation (TCOR)
Tel: +39-06-57053659
e-mail: rodrique.vinet@fao.org

FAO TECHNICAL PAPERS

FAO PLANT PRODUCTION AND PROTECTION PAPERS

1	Horticulture: a select bibliography, 1976 (E)
2	Cotton specialists and research institutions in selected countries, 1976 (E)
3	Food legumes: distribution, adaptability and biology of yield, 1977 (E F S)
4	Soybean production in the tropics, 1977 (C E F S)
4 Rev.1	Soybean production in the tropics (first revision), 1982 (E)
5	Les systèmes pastoraux sahéliens, 1977 (F)
6	Pest resistance to pesticides and crop loss assessment – Vol. 1, 1977 (E F S)
6/2	Pest resistance to pesticides and crop loss assessment – Vol. 2, 1979 (E F S)
6/3	Pest resistance to pesticides and crop loss assessment – Vol. 3, 1981 (E F S)
7	Rodent pest biology and control – Bibliography 1970–74, 1977 (E)
8	Tropical pasture seed production, 1979 (E F** S**)
9	Food legume crops: improvement and production, 1977 (E)
10	Pesticide residues in food, 1977 – Report, 1978 (E F S)
10 Rev.	Pesticide residues in food 1977 – Report, 1978 (E)
10 Sup.	Pesticide residues in food 1977 – Evaluations, 1978 (E)
11	Pesticide residues in food 1965-78 – Index and summary, 1978 (E F S)
12	Crop calendars, 1978 (E/F/S)
13	The use of FAO specifications for plant protection products, 1979 (E F S)
14	Guidelines for integrated control of rice insect pests, 1979 (Ar C E F S)
15	Pesticide residues in food 1978 – Report, 1979 (E F S)
15 Sup.	Pesticide residues in food 1978 – Evaluations, 1979 (E)
16	Rodenticides: analyses, specifications, formulations, 1979 (E F S)
17	Agrometeorological crop monitoring and forecasting, 1979 (C E F S)
18	Guidelines for integrated control of maize pests, 1979 (C E)
19	Elements of integrated control of sorghum pests, 1979 (E F S)
20	Pesticide residues in food 1979 – Report, 1980 (E F S)
20 Sup.	Pesticide residues in food 1979 – Evaluations, 1980 (E)
21	Recommended methods for measurement of pest resistance to pesticides, 1980 (E F)
22	China: multiple cropping and related crop production technology, 1980 (E)
23	China: development of olive production, 1980 (E)
24/1	Improvement and production of maize, sorghum and millet – Vol. 1. General principles, 1980 (E F)
24/2	Improvement and production of maize, sorghum and millet – Vol. 2. Breeding, agronomy and seed production, 1980 (E F)
25	*Prosopis tamarugo:* fodder tree for arid zones, 1981 (E F S)
26	Pesticide residues in food 1980 – Report, 1981 (E F S)
26 Sup.	Pesticide residues in food 1980 – Evaluations, 1981 (E)
27	Small-scale cash crop farming in South Asia, 1981 (E)
28	Second expert consultation on environmental criteria for registration of pesticides, 1981 (E F S)
29	Sesame: status and improvement, 1981 (E)
30	Palm tissue culture, 1981 (C E)
31	An eco-climatic classification of intertropical Africa, 1981 (E)
32	Weeds in tropical crops: selected abstracts, 1981 (E)
32 Sup.1	Weeds in tropical crops: review of abstracts, 1982 (E)
33	Plant collecting and herbarium development, 1981 (E)
34	Improvement of nutritional quality of food crops, 1981 (C E)
35	Date production and protection, 1982 (Ar E)
36	El cultivo y la utilización del tarwi – *Lupinus mutabilis* Sweet, 1982 (S)
37	Pesticide residues in food 1981 – Report, 1982 (E F S)
38	Winged bean production in the tropics, 1982 (E)
39	Seeds, 1982 (E/F/S)
40	Rodent control in agriculture, 1982 (Ar C E F S)
41	Rice development and rainfed rice production, 1982 (E)
42	Pesticide residues in food 1981 – Evaluations, 1982 (E)
43	Manual on mushroom cultivation, 1983 (E F)
44	Improving weed management, 1984 (E F S)
45	Pocket computers in agrometeorology, 1983 (E)
46	Pesticide residues in food 1982 – Report, 1983 (E F S)
47	The sago palm, 1983 (E F)
48	Guidelines for integrated control of cotton pests, 1983 (Ar E F S)
49	Pesticide residues in food 1982 – Evaluations, 1983 (E)
50	International plant quarantine treatment manual, 1983 (C E)
51	Handbook on jute, 1983 (E)
52	The palmyrah palm: potential and perspectives, 1983 (E)
53/1	Selected medicinal plants, 1983 (E)
54	Manual of fumigation for insect control, 1984 (C E F S)
55	Breeding for durable disease and pest resistance, 1984 (C E)
56	Pesticide residues in food 1983 – Report, 1984 (E F S)
57	Coconut, tree of life, 1984 (E S)
58	Economic guidelines for crop pest control, 1984 (E F S)
59	Micropropagation of selected rootcrops, palms, citrus and ornamental species, 1984 (E)
60	Minimum requirements for receiving and maintaining tissue culture propagating material, 1985 (E F S)
61	Pesticide residues in food 1983 – Evaluations, 1985 (E)
62	Pesticide residues in food 1984 – Report, 1985 (E F S)
63	Manual of pest control for food security reserve grain stocks, 1985 (C E)
64	Contribution à l'écologie des aphides africains, 1985 (F)
65	Amélioration de la culture irriguée du riz des petits fermiers, 1985 (F)
66	Sesame and safflower: status and potentials, 1985 (E)
67	Pesticide residues in food 1984 – Evaluations, 1985 (E)
68	Pesticide residus in food 1985 – Report, 1986 (E F S)
69	Breeding for horizontal resistance to wheat diseases, 1986 (E)
70	Breeding for durable resistance in perennial crops, 1986 (E)

71 Technical guideline on seed potato micropropagation and multiplication, 1986 (E)
72/1 Pesticide residues in food 1985 – Evaluations – Part I: Residues, 1986 (E)
72/2 Pesticide residues in food 1985 – Evaluations – Part II: Toxicology, 1986 (E)
73 Early agrometeorological crop yield assessment, 1986 (E F S)
74 Ecology and control of perennial weeds in Latin America, 1986 (E S)
75 Technical guidelines for field variety trials, 1993 (E F S)
76 Guidelines for seed exchange and plant introduction in tropical crops, 1986 (E)
77 Pesticide residues in food 1986 – Report, 1986 (E F S)
78 Pesticide residues in food 1986 – Evaluations – Part I: Residues, 1986 (E)
78/2 Pesticide residues in food 1986 – Evaluations – Part II: Toxicology, 1987 (E)
79 Tissue culture of selected tropical fruit plants, 1987 (E)
80 Improved weed management in the Near East, 1987 (E)
81 Weed science and weed control in Southeast Asia, 1987 (E)
82 Hybrid seed production of selected cereal, oil and vegetable crops, 1987 (E)
83 Litchi cultivation, 1989 (E S)
84 Pesticide residues in food 1987 – Report, 1987 (E F S)
85 Manual on the development and use of FAO specifications for plant protection products, 1987 (E** F S)
86/1 Pesticide residues in food 1987 – Evaluations – Part I: Residues, 1988 (E)
86/2 Pesticide residues in food 1987 – Evaluations – Part II: Toxicology, 1988 (E)
87 Root and tuber crops, plantains and bananas in developing countries – challenges and opportunities, 1988 (E)
88 *Jessenia* and *Oenocarpus*: neotropical oil palms worthy of domestication, 1988 (E S)
89 Vegetable production under arid and semi-arid conditions in tropical Africa, 1988 (E F)
90 Protected cultivation in the Mediterranean climate, 1990 (E F)
91 Pastures and cattle under coconuts, 1988 (E S)
92 Pesticide residues in food 1988 – Report, 1988 (E F S)
93/1 Pesticide residues in food 1988 – Evaluations – Part I: Residues, 1988 (E)
93/2 Pesticide residues in food 1988 – Evaluations – Part II: Toxicology, 1989 (E)
94 Utilization of genetic resources: suitable approaches, agronomical evaluation and use, 1989 (E)
95 Rodent pests and their control in the Near East, 1989 (E)
96 *Striga* – Improved management in Africa, 1989 (E)
97/1 Fodders for the Near East: alfalfa, 1989 (Ar E)
97/2 Fodders for the Near East: annual medic pastures, 1989 (Ar E F)
98 An annotated bibliography on rodent research in Latin America 1960-1985, 1989 (E)
99 Pesticide residues in food 1989 – Report, 1989 (E F S)
100 Pesticide residues in food 1989 – Evaluations – Part I: Residues, 1990 (E)
100/2 Pesticide residues in food 1989 – Evaluations – Part II: Toxicology, 1990 (E)
101 Soilless culture for horticultural crop production, 1990 (E)
102 Pesticide residues in food 1990 – Report, 1990 (E F S)
103/1 Pesticide residues in food 1990 – Evaluations – Part I: Residues, 1990 (E)
104 Major weeds of the Near East, 1991 (E)
105 Fundamentos teórico-prácticos del cultivo de tejidos vegetales, 1990 (S)
106 Technical guidelines for mushroom growing in the tropics, 1990 (E)
107 *Gynandropsis gynandra* (L.) Briq. – a tropical leafy vegetable – its cultivation and utilization, 1991 (E)
108 Carambola cultivation, 1993 (E S)
109 Soil solarization, 1991 (E)
110 Potato production and consumption in developing countries, 1991 (E)
111 Pesticide residues in food 1991 – Report, 1991 (E)
112 Cocoa pest and disease management in Southeast Asia and Australasia, 1992 (E)
113/1 Pesticide residues in food 1991 – Evaluations – Part I: Residues, 1991 (E)
114 Integrated pest management for protected vegetable cultivation in the Near East, 1992 (E)
115 Olive pests and their control in the Near East, 1992 (E)
116 Pesticide residues in food 1992 – Report, 1993 (E F S)
117 Quality declared seed, 1993 (E F S)
118 Pesticide residues in food 1992 – Evaluations – Part I: Residues, 1993 (E)
119 Quarantine for seed, 1993 (E)
120 Weed management for developing countries, 1993 (E S)
121 Rambutan cultivation, 1993 (E)
122 Pesticide residues in food 1993 – Report, 1993 (E F S)
123 Rodent pest management in eastern Africa, 1994 (E)
124 Pesticide residues in food 1993 – Evaluations – Part I: Residues, 1994 (E)
125 Plant quarantine: theory and practice, 1994 (Ar)
126 Tropical root and tuber crops – Production, perspectives and future prospects, 1994 (E)
127 Pesticide residues in food 1994 – Report, 1994 (E)
128 Manual on the development and use of FAO specifications for plant protection products – Fourth edition, 1995 (E F S)
129 Mangosteen cultivation, 1995 (E)
130 Post-harvest deterioration of cassava – A biotechnology perspective, 1995 (E)
131/1 Pesticide residues in food 1994 – Evaluations – Part I: Residues, Volume 1, 1995 (E)
131/2 Pesticide residues in food 1994 – Evaluations – Part I: Residues, Volume 2, 1995 (E)
132 Agro-ecology, cultivation and uses of cactus pear, 1995 (E)
133 Pesticide residues in food 1995 – Report, 1996 (E)
134 Number not assigned
135 Citrus pest problems and their control in the Near East, 1996 (E)
136 El pepino dulce y su cultivo, 1996 (S)
137 Pesticide residues in food 1995 – Evaluations – Part I: Residues, 1996 (E)
138 Sunn pests and their control in the Near East, 1996 (E)
139 Weed management in rice, 1996 (E)
140 Pesticide residues in food 1996 – Report, 1997 (E)
141 Cotton pests and their control in the Near East, 1997 (E)
142 Pesticide residues in food 1996 – Evaluations – Part I: Residues, 1997 (E)

143 Management of the whitefly-virus complex, 1997 (E)
144 Plant nematode problems and their control in the Near East region, 1997 (E)
145 Pesticide residues in food 1997 – Report, 1998 (E)
146 Pesticide residues in food 1997 – Evaluations – Part I: Residues, 1998 (E)
147 Soil solarization and integrated management of soilborne pests, 1998 (E)
148 Pesticide residues in food 1998 – Report, 1999 (E)
149 Manual on the development and use of FAO specifications for plant protection products – Fifth edition, including the new procedure, 1999 (E)
150 Restoring farmers' seed systems in disaster situations, 1999 (E)

Availability: May 1999

Ar – Arabic
C – Chinese
E – English
F – French
P – Portuguese
S – Spanish

Multil – Multilingual
* Out of print
** In preparation

The FAO Technical Papers are available through the authorized FAO Sales Agents or directly from Sales and Marketing Group, FAO, Viale delle Terme di Caracalla, 00100 Rome, Italy.